New Techniques in Tumor Localization and Radioimmunoassay

New Techniques in Tumor Localization and Radioimmunoassay

Edited by

Millard N. Croll, M.D.

Professor of Radiation Therapy and Nuclear Medicine
Hahnemann Medical College
Philadelphia, Pennsylvania

Luther W. Brady, M.D.

Professor and Chairman, Department of Radiation Therapy and Nuclear Medicine
Hahnemann Medical College
Philadelphia, Pennsylvania

Takashi Honda, M.D.

Associate Professor of Radiation Therapy and Nuclear Medicine
Hahnemann Medical College
Philadelphia, Pennsylvania

Robert J. Wallner, D.O.

Assistant Professor of Radiation Therapy and Nuclear Medicine
Hahnemann Medical College
Philadelphia, Pennsylvania

A Wiley Biomedical-Health Publication

JOHN WILEY & SONS

New York • London • Sydney • Toronto

Copyright © 1974, by John Wiley & Sons, Inc.

All rights reserved. Published simultaneously in Canada.

No part of this book may be reproduced by any means, nor transmitted, nor translated into a machine language without the written permission of the publisher.

Library of Congress Cataloging in Publication Data:

New techniques in tumor localization and radioimmunoassay.

   (A Wiley biomedical-health publication)
   Papers presented at a symposium held at the Hahnemann Medical College and Hospital, May 3–5, 1973.
   Includes bibliographical references.
   1. Tumors—Diagnosis—Congresses. 2. Radioimmunoassay—Congresses. I. Croll, Millard N., ed. II. Hahnemann Medical College and Hospital of Philadelphia. [DNLM: 1. Neoplasms—Diagnosis—Congresses. 2. Radioimmunoassay—Congresses. 3. Radioisotope scanning—Congresses. QZ241 N532 1973]

RC255.N48   616.9′92′075   74-9867
ISBN 0-471-18836-0

Printed in the United States of America

10 9 8 7 6 5 4 3 2 1

Authors

William H. Beierwaltes, M.D.

    Professor of Medicine
    Physician-in-Charge
    Division of Nuclear Medicine
    University of Michigan Medical Center
    Ann Arbor
    Michigan

Ted Bloch, M.D.

    Chief
    Nuclear Medicine Department
    Touro Infirmary
    New Orleans
    Louisiana

William H. Briner

    Captain, USPHS (Retired)
    Assistant Professor of Radiology
    Director, Section on Radiopharmacy
    Duke University Medical Center
    Durham
    North Carolina

Gerald A. Bruno, Ph.D.

    Director
    Diagnostics Research & Development
    E. R. Squibb & Sons, Inc.
    New Brunswick
    New Jersey

Helen D. Busby, R.T., N.M.T.

    Chief Technologist
    Nuclear Medicine Department
    Touro Infirmary
    New Orleans
    Louisiana

Paul L. Carmichael, M.D.

    Associate Surgeon
    Retina Service
    Wills Eye Hospital
    Philadelphia
    Pennsylvania

James J. Conway, M.D.

    Assistant Professor of Radiology
    McGaw-Medical Center of
      Northwestern University
    Attending Radiologist
    The Children's Memorial Hospital
    Chicago
    Illinois

Jay L. Federman, M.D.

    Director of Research
    Retina Service
    Wills Eye Hospital
    Philadelphia
    Pennsylvania

Theodore L. Goodfriend, M.D.

    Associate Professor
    Departments of Internal Medicine
      and Pharmacology
    School of Medicine
    University of Wisconsin
    Madison
    Wisconsin

Phillip Gorden, M.D.

    Senior Investigator
    Diabetes Section
    Clinical Endocrinology Branch

National Institute of Arthritis,
   Metabolism and Digestive Diseases
National Institutes of Health
Bethesda
Maryland

C. Craig Harris, M.S.

Associate Professor of Radiology
Division of Nuclear Medicine
Department of Radiology
Duke University Medical Center
Durham
North Carolina

Ned D. Heindel, Ph.D.

Associate Professor Chemistry
Lehigh University
Bethlehem
Pennsylvania
and
Visiting Associate Professor
Department of Radiation Therapy and
   Nuclear Medicine
Hahnemann Medical College
Philadelphia
Pennsylvania

Carla M. Hendricks

Research Technologist
Diabetes Section
Clinical Endocrinology Branch
National Institute of Arthritis,
   Metabolism and Digestive Diseases
National Institutes of Health
Bethesda
Maryland

John U. Hidalgo, M.S.

Director
Radiation Laboratory
Tulane University
New Orleans
Louisiana

Gerald C. Holst, Ph.D.

Research Physicist
Frankford Arsenal
Philadelphia
Pennsylvania

Mary Irvin, M.S.

Research Associate
Radiation Laboratory
Tulane University
New Orleans
Louisiana

John Langan, Ph.D.

Director of Research and Development
Roche Clinical Laboratories, Inc.
Raritan
New Jersey

John H. Laragh, M.D.

Columbia-Presbyterian Medical Center
Department of Medicine
Columbia University
College of Physicians and Surgeons
New York

James E. McGuigan, M.D.

Professor of Medicine
Chief, Division of Gastroenterology
University of Florida College of Medicine
Gainesville
Florida

Linn McMullin, B.S., R.T.

Technologist
Nuclear Medicine Department
Touro Infirmary
New Orleans
Louisiana

Jean E. Sealey, B.Sc.

Research Associate in Medicine
Columbia-Presbyterian Medical Center
Department of Medicine
Columbia University
College of Physicians and Surgeons
New York

Jerry A. Shields, M.D.

Director of Oncology Branch
Retina Service
Wills Eye Hospital
Philadelphia
Pennsylvania

Thomas W. Smith, M.D.

Assistant Professor of Medicine
Harvard Medical School
Associate Program Director
Myocardial Infarction Research Unit
Massachusetts General Hospital
Boston
Massachusetts

AUTHORS

Richard P. Spencer, M.D., Ph.D.

    Professor of Nuclear Medicine
    Yale University School of Medicine
    New Haven
    Connecticut

Leonard Spohrer

    Senior Programmer
    Computer Sciences
    Touro Infirmary
    New Orleans
    Louisiana

Merilyn Trenchard, B.S., R.T.

    Technologist
    Nuclear Medicine Department
    Touro Infirmary
    New Orleans
    Louisiana

Morton B. Weinstein, M.D.

    Assistant Professor of Radiology and Medicine
    University of Miami School of Medicine
    Miami
    Florida

Preface

The utilization of radionuclide techniques has revolutionized research methods, has led to a better understanding in many basic biological processes, and has produced new studies providing diagnostic information previously unobtainable.

In the field of nuclear medicine, specific organ imaging has been one of the most actively advancing research areas. This expanding interest results from the immediate clinical application of new diagnostic scanning radiopharmaceuticals in demonstrating anatomic changes in the target organ, and in assisting with positive early delineation of malignant tumors. The introduction of multiple short-lived nuclides, particularly technetium^{99m}, along with improvement in detection instrumentation, has led to successful visualization of multiple organs with a marked improvement in tumor localization.

Radioimmunoassay is a technique to measure biologically potent compounds that lack chemical groups susceptible to sensitive analysis. These highly potent biological compounds pose the greatest interest and difficulty in terms of analysis. Until the development of radioimmunoassay techniques, bioassay was the method that combined the greatest specificity and sensitivity. Instead of using an obvious physiological response of a biological system *in vivo* to both detect and measure the agents under investigation, radioimmunoassay uses the biological response of immunity to supply a sensitive and specific reagent, antibody.

The latest developments in tumor localization and radioimmunoassay were presented at a symposium, New Techniques in Tumor Localization and Radioimmunoassay, held at the Hahnemann Medical College and Hospital on May 3–5, 1973. The purpose of the symposium was to dramatize the new information available relative to tumor localization techniques and radioimmunoassay procedures, and to bridge the gap between outstanding nuclear medicine research laboratories and practicing physicians and trainees in nuclear medicine. The purpose of the symposium was to bring together prominent investigators covering these recent advances. It is the editors' firm belief that this was accomplished.

A special debt of gratitude is due to the Division of Nuclear Medicine of E. R. Squibb and Sons, Inc. for their financial support of the symposium.

The editors are grateful to the contributors for their willingness to undertake the task of commiting their comments to a written text and their patience with our persistent stimuli for completion of their manuscripts. We are appreciative of the help given by the publishers in the preparation of the volume, as well as to Mr. Carl Karsch, Ms. Virginia Preston, and Mrs. Teresa Yeykal for their able and patient editorial assistance.

MILLARD N. CROLL, M.D.
LUTHER W. BRADY, M.D.
TAKASHI HONDA, M.D.
ROBERT J. WALLNER, D.O.

March 1974

Contents

1	**Fundamentals of Radioimmunoassay** Theodore L. Goodfriend, M.D.	1
2	**Commercial Development of the Radioimmunoassay Kit** Gerald A. Bruno, Ph.D.	9
3	**The Measurement of Plasma Insulin and Proinsulinlike Components** Phillip Gorden, M.D. Carla M. Hendricks	17
4	**Measurement of Cardiac Glycoside Concentrations in Serum or Plasma: Technical Problems and Clinical Implications** Thomas W. Smith, M.D.	25
5	**Renin Activity Assay: Angiotensin I Generation and Radioimmunoassay** Jean E. Sealey, B.Sc. John H. Laragh, M.D.	39
6	**The Detection of Hepatitis B Antigen (HBAg) by Radioimmunoassay** William H. Briner, Captain, USPHS (Retired)	51
7	**The Medical Application of the Carcinoembryonic Antigen Assay** John Langan, Ph.D.	57
8	**Radioimmunoassay of Gastrin** James E. McGuigan, M.D.	65
9	**Computer Applications in Radioimmunoassay** John U. Hidalgo, M.S. Mary Irvin, M.S.	75

Leonard Spohrer
Ted Bloch, M.D.
Helen D. Busby, R.T., N.M.T.
Merilyn Trenchard, B.S., R.T.
Linn McMullin, B.S., R.T.

10 **The Chemical Foundation of Tumor Localization** 83

Ned D. Heindel, Ph.D.

11 **The Role of a Radiopharmacist in the Development of a Tumor-Localizing Radiopharmaceutical** 93

William H. Briner, Captain, USPHS (Retired)

12 **Instrumentation Factors in Visualization of Tumors** 99

C. Craig Harris, M.S.

13 **Development and Utilization of Tumor Localizing Radiopharmaceuticals (^{111}In and ^{67}Ga)** 121

Morton B. Weinstein, M.D.

14 **Labeled Chloroquine Analog in Diagnosis of Ocular and Dermal Melanomas** 161

William H. Beierwaltes, M.D.

15 **Radiolabeled Amino Acids, Antigens, and Organic Compounds in Tumor Localization** 171

Richard P. Spencer, M.D., Ph.D.

16 **Adrenal Tumor Localization with Iodocholesterol** 179

William H. Beierwaltes, M.D.

17 **The Present Status of the ^{32}P Test in Ophthalmology** 193

Paul L. Carmichael, M.D.
Gerald C. Holst, Ph.D.
Jay L. Federman, M.D.
Jerry A. Shields, M.D.

18 **The Characteristic Radionuclide Appearance of Certain Pediatric Central Nervous System Neoplasms** 203

James J. Conway, M.D.

Index 215

New Techniques in
Tumor Localization and
Radioimmunoassay

Fundamentals of Radioimmunoassay

CHAPTER ONE THEODORE L. GOODFRIEND, M.D.

RATIONALE

Radioimmunoassay was developed as a technique of measuring biologically potent compounds that lack chemical groups susceptible to sensitive analysis. Examples of biologically active compounds that do *not* require radioimmunoassay for their detection include: water, present at 55 M concentration and measured by its mass; sodium, present at 0.15 M concentration and detectable by flame photometry; albumin, present at 10^{-3} M and detectable by its mass; and hemoglobin, present in nearly saturating concentrations and detectable by its color.

By contrast, the highly potent biological compounds of greatest interest today pose new and difficult problems to analytic chemists. Insulin, for example, circulates at 10^{-10} M, and looks like a mere blob of protein to a chemist. It has virtually nothing in its overall composition to distinguish it from a dozen other polypeptides. Even if it did, there are no chemical techniques sensitive enough to detect characteristic groups or chromophores at that concentration. Biological systems, however, can detect and distinguish hormones and drugs at concentrations of 10^{-10} M or less. That fact accounts for the popularity of bioassays. They depend on the living system not only to discriminate one chemical from another, but also to react in some measurable way.

Until the development of radioimmunoassay, bioassay was the method that combined the greatest specificity and sensitivity. Instead of using an obvious physiological response of a biological system *in vivo* to both detect and measure the agents under investigation, radioimmunoassay uses the biological response to immunity to supply a sensitive and specific reagent, antibody.

The actual detection technique used in radioimmunoassay is the detection of radioactivity. The object of the assay is a *non*radioadioactive hormone or drug in human or animal samples. Therefore the technique is by definition *indirect*. Radioimmunoassay measures the effect of a biological substance on a system containing antibody and radioactive standards, neither of which are in the sample at the start.

Radioactivity can now be detected and measured with exquisite sensitivity, thanks in large part to the fact that it comes in discrete packages (quanta or disintegrations) that can be discriminated from "noise." With this technique one can detect 10^{-15} M concentrations of substances, in contrast to 10^{-8} M concentrations detectable by absorption spectroscopy, for example, Radioimmunoassay takes advantage of both the sensitivity of radiation measurements and the specificity of biological reagents. This extreme sensitivity has brought to light an esoteric vocabulary for the description of small amounts of materials: 1 nanogram = 10^{-9} grams; 1 picogram = 10^{-12} grams; and 1 femtogram = 10^{-15} grams.

REAGENTS

The necessary reagents, antibody and radioactive standard antigen, are generated by a variety of approaches, only a few of which are mentioned here. Antisera can occur naturally, can occur in patients receiving medication (as was the case in the first radioimmunoassay devised by Berson and Yalow), or can be induced in laboratory animals. When animals are immunized deliberately, the antigen is frequently modified by coupling it to carriers or by adding it to irritating adjuvants to increase the response of the immunized animal. Many factors can be altered to increase the immune response. In general, the most powerful tool for the induction of antisera is the natural diversity among animals, and the best way to take advantage of this tool is to immunize large numbers of animals and allow the best to show their genius.

The customary source of antibody is rabbits. In place of rabbits, one can use guinea pigs, horses, goats, or people. In place of antibodies, one can use binding globulins found normally in target organs, the "receptors" or in plasma, the "transport" globulins. Receptors and binding globulins are designed by evolution to react with normal circulating levels of hormones, whereas diseases usually cause levels that are much lower or higher than normal. Therefore "radioreceptor" assasys may fail to measure accurately in important ranges which can be approached by use of antisera selected for their binding range.

The second crucial ingredient for radioimmunoassay is a labeled standard. Radioactive standards can be produced by adding radioactive nuclides, commonly iodine-125. This atom combines covalently with tyrosine residues. Tyrosine residues occur in many antigens studied by radioimmunoassay, and can be added to others by a variety of techniques that do not impair the ability of the standard to react with antibody. Iodine-125 has qualities that make it superior to iodine-131, such as longer half-life, purer source, and greater counting efficiency. Other nuclides can be coupled to standards, or incorporated into them. Phosphorus-32 and tritium are useful in this regard, but less common in practice. For one thing, they must be counted by techniques more complex than the simple gamma-ray devices available for the iodines.

BASIC PROCEDURES

With the radioactive standard antigen and specific antibody, the ingredients for successful radioimmunoassay are at hand. The basic phenomenon that establishes the assay itself is the competition of a variable amount of antigen in the same and a fixed amount of radioactive standard, for a small number of specific binding sites in

BASIC PROCEDURES

the antiserum or other binding material (Figure 1). The number of binding sites must be sufficiently small to set up the competition. This competition for a limited number of binding sites gives the assay its alternative names "competitive binding assay" and "saturation kinetics assay." These names allow for the frequent use of nonimmune binding macromolecules.

Several features determine whether an assay will succeed or not. One ingredient of success is the specificity of the antibody or binder. It must bind only antigens closely related to the labeled standard. Of course, any number of othe antigen–antibody reactions can take place in the same tube. Only the reaction involving radioactive antigen will eventually be detected by the counter. So only the antibody directed at the radioactive standard must be specific.

There are in fact two sources of specificity in radioimmunoassay, the *specificity of the antiserum,* which depends on the purity of the immunizing material and the reactivity of the immune animal, and the *purity of the labeled antigen.* The presence of miscellaneous antibodies will have no effect if the radioactive antigen is so pure that it reacts with only a small group of antibody molecules.

Although the radioactive and nonradioactive antigens should both be bound by the same binding site in the antibody or binder, it is not necessary that they be identical or that the binding site see them as identical. The only real requirement is that the *unlabeled standard and the unknown in the sample have identified reactivity with binding sites.* In other words, the radioactive standard is merely a test reagent used to monitor the amount of unlabeled antigen competing for binding sites. The amount of unlabeled antigen in the sample can be judged only by com-

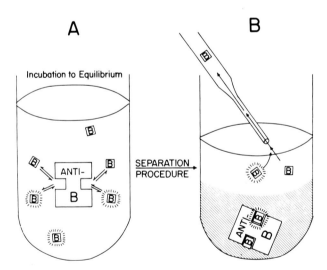

Fig. 1. The basic steps in radioimmunoassay. In the first step (*A*), a fixed amount of radioactive antigen (indicated by the radiating squares "B") and an unknown amount of nonradioactive antigen (indicated by plain squares) compete for a small number of specific binding sites on the antibody. The reaction may or may not come to equilibrium before it is terminated by the second step, a procedure that separates the binding sites and any radioactive antigen bound to them from the radioactive antigen free in solution (*B*). In this figure, the separation procedure is precipitation of binding sites by a method that leaves antigen in solution. After this procedure, the radioactivity in either the bound or free fraction (or both) is counted and serves as an indicator of the amount of nonradioactive antigen present in the original incubation mixture.

paring its effect on the radioactive reaction to standard amounts of unlabeled antigen. Therefore the standard used to build that curve, and the unknown to be measured, must react equally with the binding site. In practice, it is also advantageous to use pure, tightly bound radioactive antigen, as well as good nonradioactive standard. New tricks to purify and stabilize the label and standard have improved commercial kits, making them more homogeneous and dependable.

A second requirement for successful assay is that the antiserum be sufficiently *avid for antigen* that it binds tiny quantities. Otherwise, the sensitivity of detection of radioactivity will be wasted.

Finally, there must be a convenient way to "keep score" in the competition for the small number of binding sites. There must be a way to tell at the end of the competition how much radioactivity is bound by the sites and how much is free (Fig. 1). It is this measure that indicates, albeit indirectly, how much unlabeled antigen was present in the unknown. This "scoring system" consists of a separation step. Its goal is *rapid, complete, gentle separation* of bound from free radioactive antigen. The separation must be rapid in order to reflect accurately the status of the competition at one moment. It must be complete to be informative, and it must be gentle so as not to disrupt the equilibrium or the state of competition for binding sites.

INTERPRETATION

The results of the assay will always be: less radioactive antigen bound to binding sites when more nonradioactive antigen is present. The shape of the curve depicting these results depends, however, on whether one measures bound, free, or a ratio of bound to free radioactivity (Figure 2). The curve falls if bound radioactivity is measured, and rises if free radioactivity is measured. The predicted shape of the curve is sigmoidal when the logarithm of the amount of unlabeled material is plotted against the proportion of radioactivity bound. However, the precise curvature of the standard curve varies widely from assay to assay.

There are several conclusions regarding experimental technique to be gleaned simply from the shape of the standard curves. In the first place, values at the flat ends of the curves are less dependable, since large changes in amount of antigen cause very little change in the radioactivity bound. Second, the reproducibility of individual assays at the extremes of the curves is poorer, and the variance grows at the ends. Thus any computer transformation of data from assays of this type must include weighting factors to decrease the importance of those values, and to warn the reader that high and low values are less believable than intermediate ones.

All mathematical analyses of the standard curve assume that an equilibrium has been achieved in the reaction mixture. However, if conditions are standardized, nonequilibrium techniques can be used, and the whole assay terminated before the competition has been completed. Most investigators have found it more advisable to determine standard curves empirically, and not to depend on equilibria, mathematics, or even the law of mass action to abbreviate their efforts. The one intervention that can help is the use of computers. They enable the investigator to deal with standard curves and reaction conditions that differ from the ideal.

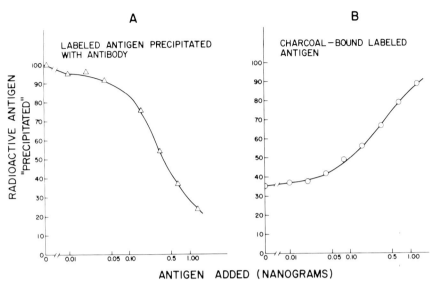

Fig. 2. Two typical standard curves of radioimmunoassay data. In each panel, radioactivity is the parameter represented by the ordinate, and unlabeled antigen by the abscissa. A is a plot of typical data from experiments such as the one depicted in Fig. 1, in which antibody-*bound* radioactivity is counted. As expected, the amount of radioactive antigen bound to antibody and precipitated with it decreases as more and more nonradioactive antigen is added to the incubation mixture. In B, the *free*, unbound radioactive antigen is trapped at the end of the incubation and precipitated with charcoal. In this type of separation step, if the charcoal pellet is counted, the amount of radioactive antigen recovered with the charcoal increases as more and more nonradioactive antigen is added. Curve B is not the precise reciprocal of curve A because some of the radioactive antigen is altered and does not obey predicted rules of binding, nonspecific adsorption can occur, the efficiency of the two procedures in recovering bound and free antigen is not identical, and their interference with the incubation equilibrium may differ.

PROBLEMS

A large number of bugs can creep into this radioimmunoassay, since it deals with biological samples, small quantities of materials, and long incubations. Figure 3 depicts how a variety of separation techniques is affected by one type of interfering substance in the reaction mixture, a substance that binds labeled antigen but is distinguishable from antibody. This artifact *increases* the apparent amount of radioactivity bound to antibody in an assay that ends with a procedure that depends on size alone. However, if the separation procedure could discriminate between antibody itself and other large molecules that bind, the artifact would show a *lower* quantity of radioactivity bound to antibody.

There are other places where artifacts can creep in: at the level of immunization, where impurities in antigens can generate antisera of high potency to materials of low interest; in the standards themselves, which may display nonhomology among species; and at the level of radioactive synthesis, where chemical reactions can do more to increase degradation than improve specific activity. Table 1 lists 10 common problems, samples of solutions, and references. There may be an unexpected rise in early points of a standard curve, attributed by some to allosteric increase in affinity overcoming competition for available sites (21).

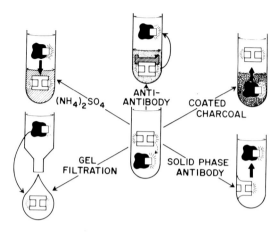

Fig. 3. Effects of nonspecific binding molecules on radioimmunoassays. The incubation, indicated in the central tube, is terminated by five different separation steps. The effects of a nonspecific, macromolecular interfering substance that binds labeled antigen are depicted for each separation step. The arrows within the tubes indicate the change in position of the radioactive antigen caused by the presence of the new binder. In the upper left corner, ammonium sulfate precipitation is shown to bring more radioactive antigen into the precipitate when a binder is present. This is interpreted as less unknown in the original mixture. In the second procedure, a specific "second antibody" is added which precipitates the first antibody. The result of a nonspecific binder is to decrease the amount of radioactivity recovered with the antibody precipitate. This is interpreted as more unknown in the original mixture. When charcoal is used to recover free antigen, binders prevent this recovery. This is interpreted as more unknown in the original mixture. Gel filtration, like ammonium sulfate, recovers all macromolecular binders regardless of whether they are antibody or not, and yields more radioactivity when a new binder is present. This is interpreted as less unknown in the original mixture. Finally, antibody bound in advance to a solid phase binds less radioactivity if there is competition by a nonspecific binder. This is interpreted as more unknown in the original mixture.

Among things that can affect results of radioimmunoassays at the incubation or competition step are such varied influences as pH, salts, protein denaturants, detergents, fragments of the antigen to be measured, enzymes that degrade the unknown or standard, adsorbent surfaces in pipettes or test tubes, and a host of factors as yet unknown. Each of these exerts unpredictable influence on the antigen–antibody reaction that forms the core of the assay. It is the unfortunate tendency of *diseased humans to differ in more than one respect* from normals. Among the differences may be changes in the above artifacts affecting assays, as well as differences in the antigen levels themselves. Therefore the construction of standard curves can be very complex.

The ideal standard curve would be diluted in a solvent identical to the unknown sample's solvent. For example, an unknown sample of plasma ideally would be compared to standard amounts of antigen diluted in the same patient's plasma. Of course this is impossible, since the patient already has some unknown antigen in his plasma, in addition to all the artifact inducers we are concerned about. There is no assurance that any technique to extract the unknown from the sample, or to render the sample free of unknown to allow additions of standard, will leave all the crucial artifacts untouched. The only real solution to this problem is the use of antisera of such high affinity and such great specificity that the immunoassay reaction will ig-

Table 1 Common Problems in Radioimmunoassay

Problem	Approach	Reference
Low titer antisera	Antigen–carrier coupling	1–3
	Modify immunization schedule, adjuvants, and sites	4
	Change strains and species	5
Low-specific-activity labeled antigen	Carrier-free tritiation; use iodine-125, not iodine-131	6 7
	Add iodinatable residue	8, 9
	Purify labeled from unlabeled	10, 11
Poor reactivity of labeled antigen with antiserum	Purify best ligand (e.g., mono- versus diiodinated)	11
Inappropriate standards	Avoid species differences	12
Interfering substances	Extract unknowns from sample, or use sample as diluent	13, 14
Incomplete bound-free separation	Alter time of exposure, or concentration of adsorbent	15–17
Disruption of equilibrium by separation	Use gentler technique (e.g., double antibody)	14
Poor sensitivity, reproducibility or accuracy	Improve pipetting technique; count more disintegrations	18
High "blank" due to adsorption, or degradation of label	Tubes or pipettes should be protein-coated, or protein and enzyme inhibitors added to buffer	5, 14, 19
"Allostery"	Use it or change antisera	20, 21

nore all potential artifact inducers and react only in relation to antigen. At this time, only a few of the radioimmunoassay kits contain such exquisitely selective antisera. The so-called nonspecific reactants in radioimmunoassay are the major impediments to its ready acceptance in routine laboratories.

In conclusion, radioimmunoassay is a technique combining the specificity of biological systems with the sensitivity of nuclear physics and engineering. It measures very small amounts of biologically active compounds. Its basic principles are as simple as the law of mass action, but the reagents are labile and susceptible to a large number of laboratory mishaps. Therefore the technique is empirical, the data interpretation must be flexible, and the role of human expertise appears to be inevitable.

ACKNOWLEDGMENTS

This presentation is dedicated to Dr. Rosalyn Yalow, and the late Dr. Sol Berson, who originated the method of radioimmunoassay, and who helped the author begin his work and then served as his helpful critic and guide. The author's experimental work was performed with Mr. Dennis Ball, and was supported by Research Grant HL-09922 from the National Institutes of Health.

REFERENCES

1. Goodfriend, T. L., Levine, L., and Fasman, G.: *Science* **144**:1344, 1964.
2. Reichlin, M., Schnure, J. J., and Vance, V. K.: *Soc. Exp. Biol. Med.* **128**:347, 1968.
3. Williams, C. A. and Chase, M. W. (eds.), *Methods in Immunology and Immunochemistry*, Vol. 1. Academic Press, New York, 1967.
4. Vaitukaitis, J., Robbins, J. B., Neischlag, E., and Ross, G. T.: *J. Clin. Endocrinol.* **33**:988, 1971.
5. Talamo, R. C., Haber, E., and Austen, K. F.: *J. Immunol.* **101**:333, 1968.
6. Morgat, J., Hung, L., and Fromageot, P.: *Biochem. Biophys. Acta* **207**:374, 1970.
7. Freedlender, A. E.: In Margoulies, M. (ed.), *Proteins and Polypeptide Hormones*, Part 2. Excerpta Medical Foundation, Amsterdam, 1968, p. 351.
8. Goodfriend, T. L. and Ball, D.: *J. Lab. Clin. Med.* **73**:501, 1969.
9. Newton, W. T., McGuigan, J. E., and Jaffe, B. M.: *J. Lab. Clin. Med.* **75**:886, 1970.
10. Glick, S. M., Kumaresan, P., Kagan, A., and Wheeler, M.: In Margoulies, M. (ed.), *Protein and Polypeptide Hormones*, Part 1. Excerpta Medical Foundation, Amsterdam, 1968, p. 81.
11. Neilsen, M. D., Jorgensen, M., and Giese, J.: *Acta Endocrinol.* **67**:104, 1971.
12. Hollemans, H. J. G.: *Clin. Chim. Acta* **27**:99, 1970.
13. Thorell, J. I.: In Margoulies M. (ed.), *Proteins and Polypeptide Hormones*, Part 2. Excerpta Medical Foundation, Amsterdam, 1968, p. 335.
14. Goodfriend, T. L., Farley, D., and Ball, D.: *J. Lab. Clin. Med.* **72**:648, 1968.
15. Waxman, S., Goodfriend, T. L., and Herbert, V.: *Clin. Res.* **15**:457, 1967.
16. Goodfriend, T. L.: In Yalow R. and Berson, S. (eds.), *Methods in Investigative and Diagnostic Endocrinology*. North Holland Publishing Company, Amsterdam, 1973, pp. 1158–1168.
17. Freedlender, A. E., Fyhrquist, F. Y., and Hollemans, H. J. G.: In Jaffe, B. M. and Behrman, H. (eds.), *Methods of Hormone Radioimmunoassay*. Academic Press, New York, in press.
18. Ekins, R., and Newman, B.: In Diczfalusy, E. (ed.), *Steroid Assay by Protein Binding*. Transactions of the Second Symposium on Steroid Assay by Protein Binding, Stockholm, 1970.
19. Goodfriend, T. L., and Odya, C.: In Jaffe, B. and Behrman, H., (eds.), *Methods of Hormone Radioimmunoassay*. Academic Press, New York, in press.
20. Matsukura, S., West, C. D., Ichikawa, Y., Jubiz, W., Harada, G., and Tyler, F. H.: *J. Lab. Clin. Med.* **77**:490, 1971.
21. Goodfriend, T. L. In Margoulies, M. and Greenwood, G. C. (eds.), *Protein and Polypeptide Hormones*. Excerpta Medica, Amsterdam, 1972, p. 287.

Commercial Development of the Radioimmunoassay Kit

CHAPTER TWO   GERALD A. BRUNO, Ph.D.

The radioimmunoassay (RIA) technique provides an enormously sensitive yet practical tool to measure minute quantities of substances of medical interest. The technique was just discovered in the mid-1950s, and has already been applied to the assay of hundreds of compounds. While the great majority of these assays are still restricted to research applications, a few have been developed on a commercial scale and put into routine clinical use. This process of commercializing the RIA technique is a vital link in the realization of the full potential of this invaluable tool.

CORPORATE COMMITMENT

The first, and perhaps the most difficult, step in the chain of events leading to a commercial RIA product, or group of products, is the decision by a corporate entity to enter the field. The decision is difficult because it requires the commitment of considerable sums of money to obtain the necessary personnel, equipment, and facilities to carry out the development program. Because most corporations are constantly exposed to sure-fire programs in which to invest their resources, all such proposals undergo careful scrutiny. Assuming that the company does have sufficient resources either to obtain additional personnel and facilities or to divert them from other programs, the corporate policy makers must obtain solid, unbiased answers to the following types of questions: (1) Is there a real medical need for this type of product? (2) How great is the need relative to competing needs (i.e., should the money be invested in antibiotic research instead)? (3) Is the program compatible with corporate strengths and goals? Do we already have, or can we develop a research and development capability? Do we have manufacturing, quality control, and distribution capability? Do we have the appropriate marketing force to promote and sell the product? (4) What kind of competition do we face? Can we do a better job than the competition?

All the above questions are asked for the obvious reason that all profit-making corporations expect a reasonable return on their investments, and try to minimize their involvement in money-losing situations. It is admittedly rare, but especially rewarding, when the profit motive can be combined with a genuine opportunity to contribute to the health and well-being of one's fellowman. We believe that the RIA field represents one of those rare opportunities.

MANPOWER, EQUIPMENT, AND FACILITIES

Once the long tedious process of obtaining corporate commitment has been completed, the pace of activities directed toward eventual marketing of a product quickens considerably. Manpower must be hired or diverted from other programs, and appropriate equipment and facilities must be made available to satisfy the unique requirements of the program.

Because of the diversity of scientific disciplines involved in RIA development, obtaining manpower for an RIA program is not a simple process. It is not possible to hire an individual with Ph.D. in radioimmunoassay, but specialized talents are required for each phase of the development program. Ideally, an organic chemist is employed to synthesize the antigen or immuogen, a radiochemist for preparation of the radioactively tagged antigen, an immunologist for immunization of animals and characterization and purification of antibodies, a pharmaceutical chemist for formulation and packaging of test reagents, and a clinical chemist to advise on the mechanics of the test procedure. In actual practice, one investigator with any of the above qualifications assumes primary responsibility for the overall assay development, and draws upon the special talents of other investigators within the organization. While there are still no individuals with Ph.D.'s in radioimmunoassay, on-the-job experience has resulted in the evolution of numerous investigators who might be readily designated as the next best thing.

The requirements for equipment and facilities are more easily satisfied, providing the corporate entity has already been engaged in development efforts employing radioisotopes and has access to conventional chemical laboratories and animal care facilities. The radioactive aspects of the development program require the appropriate radiation detection equipment (e.g., radiochromatogram scanners, well-type gamma scintillation spectrometers, liquid scintillation spectrometers, and radiation monitoring devices), and Atomic Energy Commission approved facilities for protection of laboratory personnel, including a nonrecirculating ventilation system. The facilities required for the chemical synthesis aspects of the project include the usual items associated with the synthesis and analysis of organic compounds. Depending on the type of animals used for antibody production, the animal care facilities could be as simple as a few cages to house guinea pigs, or sufficiently elaborate to accommodate a small flock of sheep or herd of goats. Considering the multitude of sample assays generated in the development program, access to automated radioactive counting equipment and electronic data processing equipment has also become a basic requirement.

PRODUCT DEVELOPMENT

With the preliminaries of funding, staffing, and housing completed, the critical phase of actually developing a product can begin. After choosing a specific assay

PRODUCT DEVELOPMENT

for development (which is usually determined by market research data generated in conjunction with obtaining a corporate commitment to the overall program), the following basic steps are involved in the product development process:

1. Obtain purified form of substances (antigen) to be assayed.
2. Prepare radioactive antigen.
3. Develop specific antibodies against the antigen.
4. Establish specific assay procedure.
5. Formulate and package test reagents.
6. Evaluate finished product in clinical laboratory.

While the above steps generally represent the order in which the project is undertaken, there is considerable overlap in their initiation and completion. Each step in the development process is discussed in greater detail.

Pure Antigen

Obtaining pure, immunologically active antigen is the critically important first step in the development process. The importance of the antigen is based on the fact that it is used for preparation of all of basic components of the assay system. The pure antigen in highly diluted solutions serves as a standard for preparation of the standard curve in the assay procedure; it is tagged with a radionuclide to serve as a tracer in the assay system; and pure or partially purified antigen is used to develop specific antibodies.

Procurement of the pure antigen can vary in complexity, from simply ordering it from a chemical supply catalog, to the time-consuming and expensive process of developing elaborate isolation and purification procedures within the laboratory. The source and cost of antigens found in several commercially available assays are tabulated below.

Antigen	Source	Cost
Digoxin	Commercial	$0.10/mg
Aldosterone	Commercial	$10/mg
Angiotensin I	Commercial	$10/mg
Gastrin	Commercial	$130/mg
Carcinoembryonic antigen	Laboratory isolation and purification	—
Australia antigen virus	Laboratory isolation and purification	—

The cost of the antigen is of little concern in the preparation of the standard and tagged antigen (where only picogram quantities are used), but can become an important consideration in the immunization program, where milligram quantities are usually required.

Radioactive Antigen

The radionuclides most commonly used for tagging the antigen are iodine-125 (^{125}I) and hydrogen-3 (^3H). Where it is possible to incorporate iodine into the compound,

^{125}I is the radioactive label of choice, for reasons of greater stability and convenience in detection. The radioiodination procedure that has evolved as the method of choice is the classic Hunter–Greenwood technique employing chloramine-T (1). The reaction is carried out on a very small scale to achieve the desired specific activities. In a typical reaction, 5 μg of the antigen, in a volume of 10 μl, is mixed with 1 mCi of ^{125}I (10 μl) and 10 μl of 0.5 M sodium phosphate. The iodination is achieved by the injection of 10 μl of chloramine-T (50 μg) into a stoppered vessel containing the above reactants. The stoppered vessel is used to avoid the release of gaseous radioactive iodine, and the volume is kept to a minimum to maintain a reasonably high concentration of the antigen, which in turn results in a reasonably high percent incorporation of the radioiodine.

The major concern associated with the tagged antigen is to achieve a specific activity (microcuries per microgram) that is sufficiently high to not affect the sensitivity of the test, but not so great that the immunological activity of the compound is altered by the presence of an excessive number of iodine atoms. Of equal concern is the purification of the tagged antigen to remove unreacted radioiodine, chemical reactants, and damaged radioiodinated antigen. Gel filtration, adsorption chromatography, and gel electrophoresis have been successfully employed in such clean-up procedures.

Antibodies

Perhaps the most difficult and time-consuming aspect of the development project is the preparation of sensitive and highly specific antibodies against the substance (antigen) to be assayed. This difficulty is, for the most part, encountered when the antigen is a relatively small molecule that is not in itself immunogenic. The problem has been partially resolved through the use of antigen "carriers" which may or may not be immunogenic. Commonly used carriers include poly- L -lysine, rabbit serum albumin, and bovine serum albumin. The antigen coupled to a carrier molecule is referred to as the immunogen.

Since the immunogen alone usually does not elicit a strong antibody response, it has become common practice to mix the immunogen with an appropriate "adjuvant" to enhance the immune response (2). The most widely used adjuvant is Freund's complete adjuvant, consisting of Arlacel A, paraffin oil, and killed mycobacteria. Another approach that has been used successfully involves adsorption of the immunogen on microparticles of carbon, polystyrene latex, polyacrylamide beads, or aluminum hydroxide. It is attributable to phagocytosis of the immunogen-laden solid material.

The choice of animal and route of administration for the immunization program are highly subjective matters. Rabbits and guinea pigs are the most widely used animals, but sheep and goats have in certain instances produced a better antibody response. It is likewise claimed that the best responses are obtained with routes of administration in the following order of superiority: lymph nodes, intraarticular, intradermal, intramuscular, intraperitoneal, subcutaneous, and intravenous, (2). As with the choice of animal species, the optimum route of administration is still an open question which appears to vary with the particular immunogen and immunization protocol being followed.

PRODUCT DEVELOPMENT

The immunogen dose and immunization schedule are also somewhat ill-defined subjects which vary with the individual investigator. An initial antigen dose of approximately 1 mg per animal is generally used, with lower dose booster injections being administered at 4-6 week intervals. The animals are usually bled 7-14 days after the booster injections to evaluate the quality of the antiserum being produced. Cardiac puncture is the commonly used technique for bleeding the smaller animal species, but ear vein bleeding in rabbits has proved to be a safer technique, especially when bleeding the better producers in the colony. In most instances, it requires from 6 to 12 months to produce high-titer, high-avidity antiserum.

The characteristics that are most critical in establishing the quality of antiserum are avidity and specificity. Titer is important from the viewpoint of the number of tests that can be derived per milliliter of antiserum harvested, but is not of major importance in terms of the sensitivity or accuracy of the final test. It should be noted that titer is defined as "the final dilution of the antiserum in the incubation mixture required to bind the appropriate amount of labelled antigen in the absence of unlabelled antigen." Useful antiserums can have titers ranging from less than a thousand to over a million, and in the case in which the titer is a million, each milliliter of the antiserum allows performance of a million tests.

The avidity of the antibody is defined as "the energy of reaction between the combing sites of an antibody and its specific antigen (antigenic determinants)," and is synonymous with sensitivity. The avidity of the antibody, and in turn the sensitivity of a particular assay system, is reflected in the slope of the standard curve obtained in the assay.

A highly specific antibody is one that does not cross-react with substances of similar structure that might be present in the sample being analyzed. The specificity of the antibody is of great importance to the accuracy of the assay procedure, especially in cases in which substances with similar structure are present in large and variable amounts.

Assay Procedure

The major concern in development of the assay procedure is the choice of a means to separate the free antigen from antibody-bound antigen. A great variety of separation techniques is available, the more prominent of which are:

Activated charcoal.
Talc.
Organic resins.
Electrophoresis.
Ammonium sulfate precipitation.
Double antibody.
Solid phase.

The activated charcoal technique (3) is probably the most widely used separation technique, but because of greater convenience for the user, the solid-phase techniques will no doubt assume great prominence in the future.

Formulation and Packaging

Perhaps the area in which the pharmaceutical manufacturer has made the most significant contribution to the commercial development of radioimmunoassay is in the formation and packaging of test reagents. Unlike research assays, the commercial assay must be formulated in such a manner that the reagents will remain stable over a reasonable period of storage in both the manufacturer's and the user's laboratory. The requirements for storage and shipment of reagents also demand that suitable packaging components be developed that will not interfere with test results.

Pharmaceutical formulation experience associated with the use of buffers, stabilizers, and preservatives, and an understanding of the characteristics of container and closure materials, has been invaluable in the development of stable reagent components.

Clinical Evaluation

Upon successful completion of the above activities, the finished product is sent to outside investigators to, hopefully, confirm the validity of the test procedure in unfamiliar hands. The basic aims of the clinical evaluation are: (a) to establish that the test consistently gives accurate results; (b) to establish normal values for this specific assay procedure; and (c) to obtain clinical values for abnormal conditions being studied.

MANUFACTURE, CONTROL, AND DISTRIBUTION

Because of the great complexity associated with the preparation and testing of the RIA reagents, and the relatively modest volume of use of such products, we have taken the approach of producing these products under the direction of the research and development personnel responsible for their development. No doubt, as many of the uncertainties associated with these preparations are eliminated, and as methodology is simplified, the production will become more automated, and the user will benefit from the economies effected.

Because of the highly dilute and sensitive nature of the more important test reagents, the products are stored and shipped at $-15°C$. To maintain this temperature during shipping, the product is packed in Dry Ice and delivered to the user within 48 hours of shipment, through use of the sophisticated, but expensive, distribution system employed for shipment of shorter-lived radiopharmaceuticals.

MARKETING

There would be little point to commercial development of RIA products were it not for the fact that eventually someone would buy the products produced, but like the other aspects of commercial RIA development, the marketing activity does not follow conventional patterns. The education of the referring physician is an extremely important factor in realizing the full potential of these assay procedures. Because the information obtainable by RIA techniques was unheard of when the majority of our practicing physicians were in training, it will take a concerted effort on the part of both the commercial manufacturer and the laboratory user to bring these assays to the attention of physicians who will request the tests.

In the context of an overall conclusion, it can be stated that the commercial development of RIA procedures is indeed a complex and at times very frustrating process. However, the negative aspects are heavily outweighed by the opportunity to play a role in the growth of this exciting new field. We have high expectations for the future of this assay procedure.

REFERENCES

1. Hunter, W. M. and Greenwood, F. C.: Preparation of iodine-131 labelled growth hormone of high specific activity. *Nature* **194**:495-496, 1962.
2. Kirkham, K. E. and Hunter, W. M.: *Radioimmunoassay Methods.* Churchill Livingstone, London, 1971, pp. 121-142.
3. Herbert, V., Gottlieb, C., Lau, K. D., and Wasserman, L. R.: Intrinsic factor assay. *Lancet* **2**:1017-1018, 1964.

The Measurement of Plasma Insulin and Proinsulinlike Components

CHAPTER THREE

PHILLIP GORDEN, M.D.
CARLA M. HENDRICKS

CLINICAL USE OF THE INSULIN RADIOIMMUNOASSAY

Radioimmunoassays, as developed by Yalow and Berson, provide a powerful tool for measurement of minute concentrations of substances present in biological fluids (1); measurement of plasma insulin has been one of the most important applications of this technique. The insulin radioimmunoassay has been applied to multiple clinical problems. Although there is still much to be learned, certain generalizations can be made about the usefulness of determining plasma insulin concentrations in the management of individual clinical problems. The physician who orders the test and the laboratory personnel who perform it must be aware of several relatively simple principles which have evolved from the vast clinical and laboratory investigations carried out since the introduction of the insulin assay.

Although multiple factors modulate insulin secretion, the most important physiological regulator is the blood glucose concentration. No interpretation can be given to a plasma insulin value unless the blood glucose concentration is known. The measurement of plasma insulin is of little diagnostic value in glucose-intolerant or hyperglycemic disorders. For example, maturity-onset diabetes is characterized by glucose intolerance, a delay in the initial phase of insulin release, and relative hyperinsulinemia occurring in the second hour after glucose administration (1–3). Though characteristic of maturity-onset diabetes, this pattern of insulin response is not unique, and may be seen in other clinical states that exhibit moderate glucose intolerance. Thus obesity, acromegaly, glucocorticoid excess, carbohydrate deprivation, and many other disorders may be associated with glucose intolerance and delayed hyperinsulinemia. Juvenile-onset diabetes, however, is characterized by severe glucose intolerance and hypoinsulinemia; other clinical disorders, such as chronic pancreatitis, cystic fibrosis, hypokalemia, or pheochromocytoma may, however, show similar hypoinsulinemic responses.

The plasma insulin response in hyperglycemic states reflects, predominantly, the severity of glucose intolerance rather than the specific etiology of the glucose intolerance. Thus while measurements of plasma insulin have contributed enormously to our understanding of physiological principles, these determinations are of limited usefulness in either diagnosis or therapy of individual patients with hyperglycemic disorders.

By contrast, the measurement of plasma insulin is indispensable in the differential diagnosis and management of spontaneous hypoglycemic disorders. The failure of insulin secretion to turn off with hypoglycemia (defined as blood glucose of 40 mg/100 ml or less) uniquely distinguishes the insulin-secreting tumor (4-6). Inappropriate insulin secretion is best demonstrated by a carefully controlled fast for periods up to 72 hours. A normal overnight fasting insulin concentration (5-25 μU/ml) becomes inappropriately high when the blood glucose concentration drops below 40 mg/100 ml. Failure to suppress plasma insulin to essentially undetectable levels in the presence of hypoglycemia suggests the diagnosis of an insulin-secreting tumor. The demonstration of inappropriately high plasma insulin concentrations with fasting hypoglycemia is essential to establish the diagnosis of islet cell tumor. When insulin concentrations are measured with brief periods of fasting, care must be taken to distinguish reactive insulin secretion that results from a meal or other stimulus. This can usually be done by frequent sequential measurements. It follows therefore that insulin measurements are of no diagnostic value in reactive or alimentary hypoglycemia, since the insulin concentrations are simply a function of the glucose tolerance curve.

Though glucose is the major physiological regulator of insulin secretion, patients who harbor islet cell tumors tend to respond less to glucose than other insulin secretagogues. Owing to this phenomenon, a series of stimulatory tests has been proposed as an adjunct to demonstrating inappropriate insulin secretion. These include measurement of plasma insulin concentrations following the administration of tolbutamide, leucine, glucagon, and calcium (4,6). These tests may be useful in confirming the diagnosis and in the follow-up evaluation of patients, but should not be used as a primary basis for a diagnostic and therapeutic decision.

Technique of Insulin Radioimmunoassay

In a typical assay, the following reagents are combined: an albumin-containing buffered solution; iodinated insulin (insulin is iodinated, usually with $Na^{125}I$ at a specific activity of 100-200 $\mu Ci/\mu g$ (7,8)); a 1:10 or greater dilution of plasma, an appropriate dilution of an extract, or standard concentrations of insulin to construct a standard curve; and guinea pig antiinsulin sera at an appropriate dilution. These reagents are mixed and incubated at 4°C for approximately 72 hours to allow the reactions to reach equilibrium. At the end of the incubation period the antibody-bound [^{125}I]insulin and free [^{125}I]insulin are separated (1). Since bound hormone does not spontaneously precipitate, some additional method must be used to separate the bound and free radioactive insulin. This can be done by immunoprecipitation (double-antibody method) (9), by adsorption of the free [^{125}I]insulin to talc (10), to charcoal (11), to cellulose (1), to other surfaces, or by a variety of gel filtration or salting-out techniques. Following separation of the bound and free labeled hormone, a bound/free ratio is calculated (B/F [^{125}I]insulin) and plotted as a function of increasing concentrations of unlabeled insulin in the standard solutions. The B/F is calculated for each unknown sample, and the immunoreactive insulin concentration read directly from the standard curve.

The integrity of each reagent is very important in achieving a high-quality assay. Bacterial growth, alterations of pH or of salt concentrations in the buffer, poor-quality antibody, or an incorrect dilution can all profoundly affect the assay. The most common problem, however, is poor-quality or low-specific-activity labeled insulin.

Other problems may be associated with a particular patient's plasma. Patients previously treated with commercial insulin or who have taken insulin surreptitiously have circulating antihuman insulin antibodies which give a false high insulin value in a double-antibody system and a false low value in an adsorption system. This can be a source of grievous error, and the physician and laboratory personnel must be alert to this possibility. Other problems that occasionally arise are damage to the labeled hormone by certain plasma samples, or contamination of the plasma, glassware, or reagents with either cold or labeled insulin or other interfering substances.

Clinical utility and general techniques involved in measuring total immunoreactive insulin have been discussed; we now describe in greater detail how insulin radioimmunoassay can be used to measure individual plasma insulin components.

Insulin and Proinsulinlike Components

Plasma immunoreactive insulin comprises at least two distinct components. One component (insulinlike component) is indistinguishable from crystalline pancreatic insulin in its gel filtration, immunological, and biological properties; the second component more closely resembles proinsulin, the biosynthetic precursor of pancreatic insulin (13), and is proinsulinlike in its gel filtration, immunological and biological properties (14) (Fig. 1). Standard radioimmunoassay procedures, as described above, measure total plasma immunoreactive insulin (1), which includes both the insulin and proinsulinlike components.

We present, here, a method of separating the individual components in plasma; with these methods, quantitative measurements can be made of the insulinlike component ("little" insulin), proinsulinlike component ("big" insulin), and the total immunoreactive insulin (12).

Collection of Specimen

For most studies 5–10 ml of blood is drawn into a syringe that had been rinsed with heparin. The plasma is separated by centrifugation (2000 rpm for 10 minutes) and stored in plastic or glass tubes at $-20°C$. No special precautions are necessary to keep the samples cold before and during centrifugation, as no conversion of the proinsulinlike component occurs in samples left at room temperature for several hours, or after 2–3 years of storage at $-20°C$. Though we routinely use heparin, other anticoagulants are equally satisfactory, as is serum.

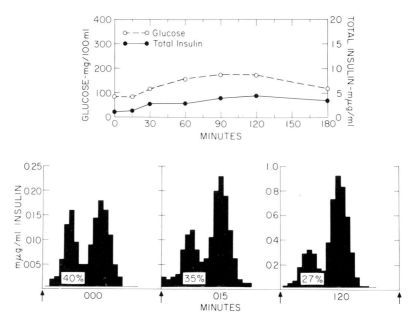

Fig. 1. Oral glucose tolerance test (100 g) in obese hypokalemic subject. Top panel represents glucose and total immunoreactive insulin plotted as a function of time after glucose administration. Lower panel represents typical G-50 (fine) Sephadex gel filtration pattern carried out in the basal state (000) and at intervals after glucose administration. The arrow to the left of each panel represents the void elution marker, and the arrow to the right the salt elution marker. (Note that in the second and third panels the void elution marker is superimposed on the salt marker of the previous column.) The immunoreactive insulin concentration is plotted as a function of the elution volume of the fraction. The peak eluting midway between the arrows is the insulinlike component, whereas the peak designated by the percent values is the proinsulinlike component and shows the percentage of the total immunoreactive insulin represented by the proinsulinlike component.

Gel Filtration Procedure

Sephadex G-50 fine* is mixed with distilled water to swell and define the beads. The beads are swollen upon exposure to water, and the fines are removed by decanting the supernatant material. Sephadex, prepared in this way, may be stored at 4°C with a small amount of some bacteriostatic agent such as 0.02% sodium azide.†

For routine studies we use a column 0.9×60 cm. Ordinary burets may be used for columns, but we have found the commercially available Pharmacia columns to be the most satisfactory for maintaining constant flow and pressure. The swollen, defined Sephadex beads are poured into the column to a height of approximately 50–55 cm, and the remaining space filled with the eluting buffer (diluent used in the radioimmunoassay (12)). Since the samples will be run at 4°C and the column must be perfectly aligned on its vertical axis, the column should be poured at its permanent station. The aspirator bottle or Mariotte flask is connected to the

---

* Pharmacia Fine Chemicals, Uppsala, Sweden.
† See *Sephadex Gel Filtration in Theory and Practice*—available from Pharmacia Fine Chemicals, Uppsala, Sweden.

column, and the column equilibrated with the eluting buffer. The rate of flow is primarily determined by the vertical distance between the outlet from the column and the reservoir bottle. This distance should be only 20–30 cm, since good resolution of the sample depends on a slow filtration rate (12–18 hours per sample). An automated fraction collector is essential for the collection of equal aliquots of the column eluate.

The plasma sample to be filtered is thawed and centrifuged to remove fibrin debris. From 0.5 to 2 ml of plasma is enriched with a trace amount of [^{125}I]albumin and ^{125}I, which mark the void and salt elution volumes of the column, respectively (12). The volume of plasma sample used is determined by the concentration of total immunoreactive insulin in the sample. The minimal insulin concentration of a workable sample for this procedure is about 1 ng/ml, and the maximum is about 10 ng/ml. The smallest volume applied is most desirable, since it gives the best resolution; samples in volumes greater than 2 ml give poor resolution on a column only 0.9×60 cm.

The sample is now applied to the Sephadex column, with care being taken not to disturb the Sephadex bed. Fractions of 0.8 ml are collected in 10×75 mm disposable tubes. The void volume of the column is approximately 12 ml, and the volume between [^{125}I]albumin and ^{125}I is approximately 20 ml. The void and salt locations are determined by counting ^{125}I, and the counts are then plotted on arithmetic graph paper, aligning the salt peak 10 units to the right of the albumin peak. Since proinsulin normally aligns 2.5 units to the right of the albumin peak and insulin 5.0 units to the right (or midway between the albumin and salt peaks), all tubes located in the 0.5–6.5 unit range should be assayed. Since the fractions were eluted in the assay diluent, they can be assayed directly, or frozen and stored.

The Assay

Routine assay procedures are now followed; that is, the 0.8 ml fractions are mixed with 0.1 ml of diluent containing porcine [^{125}I]insulin ^{125}I and carrier guinea pig plasma, followed by the addition of 0.1 ml of dilute guinea pig antiporcine insulin sera. The tubes are then incubated for 72 hours at 4°C, after which time 0.05 ml of rabbit antiguinea pig plasma is added and the incubation continued for an additional 8–18 hours. The tubes are then centrifuged at 4°C (2800 rpm for 20 minutes), the supernatant is decanted, and the radioactivity in the precipitate (bound [^{125}I]insulin) and in the supernatant (free [^{125}I]insulin) is determined by counting in a suitable gamma detector (12).

By the use of a porcine insulin standard and an antiinsulin serum that reacts similarly with the insulin and proinsulinlike components, the immunoreactive insulin concentration of each fraction is determined from the bound ^{125}I/free ^{125}I of the individual fractions assayed from the column (15). The insulin concentrations are now plotted on the vertical axis of the graph paper where the column fractions were previously aligned on the horizontal axis. A well-resolved column shows at least two distinct peaks, with the division between peaks usually occurring about 3.6 units to the right of the [^{125}I]albumin marker.

Definitions of Quantitative Measurements (15)

Total insulin concentration is the immunoreactive insulin concentration of dilute, unfiltered plasma or serum; percent proinsulinlike component is the sum of the insulin concentration of fractions in the proinsulinlike peak divided by the sum of all the fractions in both major peaks; concentration of proinsulinlike component is the percentage of proinsulinlike component times the concentration of the total insulin.

Low Insulin Concentration and Preparative Yields

Larger columns and a volatile eluant are used for samples with low insulin concentrations, for preparative yields, and for better resolution of a given sample. Thus larger amounts of plasma can be applied to these columns; the larger fractions are frozen, lyophilized thoroughly, and reconstituted in assay diluent to a workable volume.

A plasma sample of 4–12 ml containing less than 1 ng/ml can be applied to a 1.5 × 90 cm Sephadex G-50 fine column and eluted with 0.05 M $(NH_4)CO_3$, pH 8.6 (16). Fractions of 2 ml are collected, frozen, lyophilized to dryness, and reconstituted in 0.8 ml of assay diluent. This same procedure can be followed for samples with higher insulin concentrations if smaller fractions are collected and lyophilized. When dealing with samples of very high concentration (such as pooled preparative yields), aliquots of up to 0.1 ml of each fraction can be added directly to the assay.

For preparative yields of separated insulin and proinsulinlike pools on a given patient, we use a 5.0 × 100 cm column to which a 100–200 ml sample of plasma can be applied. Fractions of 12–15 ml are collected, the albumin and salt markers aligned as previously described, and the appropriate fractions pooled and lyophilized for later application to the 1.5 × 90 cm column for a final, more discrete separation of the components.

Possible Problems

For various reasons, gel filtration patterns do not always resolve satisfactorily. There may be one or more spurious peaks; the proinsulinlike material may appear more as a shoulder of the insulin component rather than as a distinct peak of its own (15); or the sampled fractions may fail to return to a base value. Whatever the peculiarity, it is best to repeat the sample with some possible modification of technique, such as applying less sample to the column, choosing a taller, thinner column, or clearing the plasma of accumulated debris and fat by recentrifuging or filtering prior to column application. Some plasmas are more viscous than others and fail to give good resolution of the two components. Also, when the percentage of the proinsulinlike material is greater than that of the insulin component, it is usually necessary to run the plasma sample on the taller column and to collect smaller fractions (16). This latter procedure has the added advantages of possibly detecting a preproinsulinlike peak in samples containing a high enough proinsulinlike concentration (17).

Clinical Utility of Fractionating Plasma Compounds

The clinical utility of separating the individual components is not yet firmly established, but may become important in the next several years. Several generalizations, however, can be made from the available data. The percent of proinsulinlike component is always highest in the basal state; stimulation of insulin secretion with glucose or other secretagogues produces primarily release of the insulinlike component, which results in a fall in the percent of proinsulinlike component. These factors must be considered in choosing samples for comparison and analysis (16). Patients with islet cell tumors tend to have a high percentage of the proinsulinlike component, but the range of values is large and overlaps with nontumor subjects (6) (Figure 2). Measurement of the proinsulinlike component may provide a useful tool for following patients with islet cell carcinomas who are treated with chemotherapeutic agents, and possibly for studying changes in islet cell function.

Although the presence of insulin antibody precludes the measurement of insulin components by the techniques described here, specific assays for the connecting peptide (the polypeptide chain in proinsulin that connects the A and B chains of insulin) can be used to measure proinsulin-related components or the free connecting

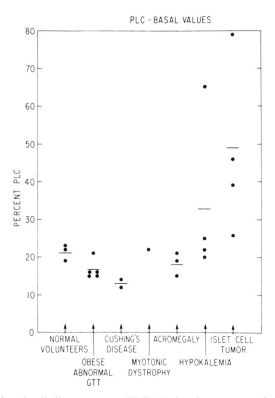

Fig. 2. Percent basal proinsulinlike component (PLC). Each point represents the basal percent proinsulinlike component of individual patients, and the bar indicates the mean for each group. The hypokalemic patients had low serum potassium values at the time of the study. (Reprinted *J. Clin. Invest.* 50:2113–2122, 1971.)

peptide (C peptide). Using these techniques, residual islet cell function can be evaluated in patients receiving exogenous insulin therapy (18) (19).

REFERENCES

1. Yalow, R. S. and Berson, S. A.: Immunoassay of endogenous plasma insulin in man. *J. Clin. Invest.* **39**:1157–1175, 1960.
2. Berson, S. A. and Yalow, R. S.: Plasma insulin. In Ellenberg, E. and Rifkin H. (eds.), *Diabetes Mellitus: Theory and Practice.* McGraw-Hill Book Company, New York, 1970, pp. 308–367.
3. Mirsky, I. A.: Regulation of secretion. In Berson, S. A. and Yalow, R. S. (eds.), *Methods in Investigative and Diagnostic Endocrinology; Part III Non-Pituitary Hormones.* North Holland Publishing Company, Amsterdam, 1973, pp. 1238–1298.
4. Roth, J. and Gorden, P.: Clinical applications of the insulin assay. In Berson, S. A. and Yalow, R. S. (eds.), *Methods in Investigative and Diagnostic Endocrinology; Part III Non-Pituitary Hormones.* North Holland Publishing Company, Amsterdam, 1973, pp. 876–883.
5. Yalow, R. S. and Berson, S. A.: Dynamics of insulin secretion in hypoglycemia. *Diabetes* **14**:341–349, 1965.
6. Schein, P. S., De Lellis, R. A., Kahn, C. R. Gorden, P., Kraft, A. R., Islet cell tumors: Current concepts and management. *Ann. Intern. Med.*, **79**:239–257, 1973.
7. Hunter, W. M. and Greenwood, F. C.: Preparation of iodine-131 labelled human growth hormone of high specific activity. *Nature* **194**:495–496, 1962.
8. Gavin, J. R., III, Roth, J. Jen, P. and Freychet, P.: Insulin receptors in human circulating cells and fibroblasts. *Proc. Nat. Acad. Sci. U.S.* **69**:747–751, 1972.
9. Morgan, C. R. and Lazarow, A.: Immunoassay of insulin: Two antibody system. Plasma insulin levels of normal, subdiabetic, and Diabetic Rats. *Diabetes* **12**:115–126, 1963.
10. Rosselin, G., Assan, R., Yalow, R. S., and Berson, S. A.: Separation of antibody-bound and unbound peptide hormones labelled with iodine-131 by talcum powder and precipitated silica. *Nature* **212**:355–358, 1966.
11. Herbert, V., Lau, K.-S., Gottlieb, C. W., and Bleicher, S. J.: Coated charcoal immunoassay of insulin. *J. Clin. Endocrinol. Metabol.* **25**:1375–1384, 1965.
12. Roth, J., Gorden, P., and Pastan, I.: "Big insulin": A new component of plasma insulin detected by immunoassay. *Proc. Nat. Acad. Sci. U.S.* **61**:138–145, 1968.
13. Steiner, D. F. and Oyer, P. E.: The biosynthesis of insulin and a probable precursor of insulin by a human islet cell adenoma. *Proc. Nat. Acad. Sci. U.S.* **57**:473–480, 1967.
14. Sherman, B. M., Gorden, P., Roth, J., and Freychet, P.: Circulating insulin: The proinsulin-like properties of "big" insulin in patients without iselt cell tumors. *J. Clin. Invest.* **50**:849–858, 1971.
15. Gorden, P. and Roth, J.: Plasma insulin: Fluctuations in the "big" insulin component in man following glucose and other stimuli. *J. Clin. Invest.* **48**:2225–2234, 1969.
16. Gorden, P., Sherman, B., and Roth, J.: Proinsulin-like component of circulating insulin in the basal state and in patients and hamsters with islet cell tumors. *J. Clin. Invest.* **50**:2113–2122, 1971.
17. Gorden, P., Freychet, P., and Nankin, H.: A unique form of circulating insulin in human islet cell carcinoma. *J. Clin. Endocrinol. Metabol.* **33**:983–987, 1971.
18. Melani, F., Rubenstein, A. H., Oyer, P. E., and Steiner, D. F.: Identification of proinsulin and C-peptide in human serum by a specific immuoassay. *Proc. Nat. Acad. Sci. U.S.* **67**:148–155, 1970.
19. Block, M. B., Rosenfield, R. L., Mako, M. E., Steiner, D. F., and Rubenstein, A. H.: Sequential changes in beta-cell function in insulin-treated diabetic patients assessed by C-peptide immunoreactivity. *N. Engl. J. Med.* **288**:1144–1148, 1973.

Measurement of Cardiac Glycoside Concentrations in Serum or Plasma: Technical Problems and Clinical Implications

CHAPTER FOUR THOMAS W. SMITH, M.D.

Digitalis intoxication occurs commonly in clinical practice (1), and this has led numerous investigators to seek better techniques for the measurement of serum or plasma cardiac glycoside concentrations. Radioimmunoassay has been a useful approach to this problem, and we discuss both technical aspects and clinical use of cardiac glycoside radioimmunoassay.

DEVELOPMENT OF METHODS FOR CARDIAC GLYCOSIDE ASSAY

Techniques in current use for the measurement of clinically relevant concentrations of cardiac glycosides in biological fluids have been recently reviewed in detail (2). To recapitulate briefly, it became possible in the 1950's to label cardiac glycosides with ^{14}C and ^{3}H. Use of these labeled tracers in experimental animals and human volunteers has provided the basis for most of our current understanding of the pharmacokinetics of digoxin and digitoxin (3). The availability of these labeled tracer compounds also provided a necessary element in most of the assay techniques subsequently developed. This is true of the double-isotope dilution derivative assay for digitoxin developed by Lukas and Peterson, which has been useful in measurements of digitoxin in plasma, whole blood, urine, and stool (4). Quantification of recovery by the use of tracer techniques is also a necessary element in the gas–liquid chromatographic method for assay of digoxin developed by Watson and Kalman (5). Both of these techniques are technically demanding, but provide high specificity and are useful in the detailed definition of metabolic pathways of cardiac glycosides.

Portions of the work described were supported by USPHS-NHLI Grant #HL 14325.

Schatzmann's discovery that cardiac glycosides are potent and specific inhibitors of transmembrane sodium and potassium transport led to the work of Lowenstein and Corrill, who developed a red blood cell ^{86}Rb uptake inhibition technique for cardiac glycoside assay (6). This approach has been modified by Grahame-Smith and Everest (7), Bertler and Redfors (8) and Gjerdrum (9). Serum digoxin and digitoxin concentrations have also been measured by direct determination of the extent to which these agents inhibit the transport enzyme Na^+,K^+-ATPase (10,11).

The techniques currently in most widespread use for determination of digitalis glycoside concentrations in serum and plasma are competitive protein binding assays of the radioimmunoassay (12,13) or enzymic displacement (14) type. Competitive binding to Na^+,K^+-ATPase has been used by Brooker and Jelliffe for the determination of digoxin and digitoxin concentrations, and their reported results are in excellent agreement with those obtained by radioimmunoassay (14). With the exception of radioimmunoassay, all the methods mentioned above require extraction of cardiac glycosides with water-immiscible organic solvents prior to measurement.

Our own experience has been based on the development of the radioimmunoassay approach. Butler and Chen immunized rabbits with conjugates of digoxin and serum albumin, and demonstrated the formation of antibodies that bind digoxin (15). Antiserums obtained in this way have been characterized in detail in terms of affinity and specificity (16). The exceptionally high affinity and specificity of selected antiserums for cardiac glycoside haptens allow quantification of subnanogram amounts of digoxin or digitoxin by convenient methods which can be utilized in the well-equipped clinical chemistry laboratory.

As outlined schematically in Figure 1 for the case of digoxin, an aliquot of serum or plasma containing unlabeled digoxin, without prior extraction, is mixed with a suitable tracer quantity of digoxin in a convenient buffer volume. The amount of tracer is chosen to lie near the midpoint of the range of unknown concentrations to be measured. This amount of tracer in turn determines the amount of antibody to

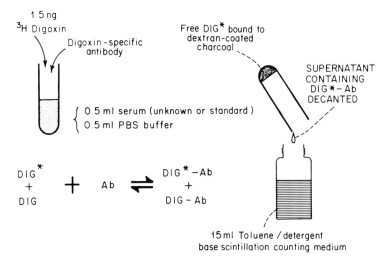

Fig. 1. Schematic representation of cardiac glycoside radioimmunoassay procedure using digoxin as an example. DIG, Digoxin; DIG*, [^{3}H]digoxin; Ab, digoxin-specific antibody.

be used. Optimal results are obtained when the percentage of total tracer bound by antibody is in the 37–50% range in the absence of any competing unlabeled ligand. The high specific activity of commercially available tritiated digoxin and the high affinity of properly selected digoxin-specific antiserums result in a sensitivity of 0.1 ng/ml or less. After an incubation period during which equilibrium is approached between tritiated and unlabeled digoxin and antibody binding sites, dextran-coated charcoal is added to achieve separation of antibody bound from free tritiated digoxin. It is very important to expose both known samples used to construct the standard curve and unknown samples to dextran-coated charcoal for similar time periods, since the equilibrium shown in Figure 1 tends to be pulled toward the left as free digoxin is bound to charcoal. This problem is of little consequence when very high-affinity antibodies such as those just described are used, but becomes an increasingly important problem with lower-affinity antiserums, as discussed below. After removal of charcoal by centrifugation, antibody-bound tritiated digoxin in the supernatant phase is decanted into liquid scintillation counting medium and counted in a liquid scintillation spectrometer. A ^{226}Ra external standard is generally used for quenching correction in our laboratory. Internal standardization is also quite satisfactory, especially if excessively large numbers of samples are not being run.

Standard curves can be plotted in several ways, including the two shown in Figure 2. We usually use the reciprocal plot shown on the right because of its applicability with computer programs (17). The computer plots the best fit for rectilinear standard curves by least-squares linear regression analysis, and compares cor-

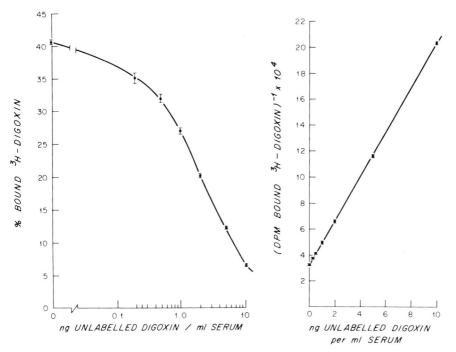

Fig. 2. Standard curves for digoxin radioimmunoassay. The same data are plotted on the left as percent antibody-bound [^{3}H]digoxin versus log unlabeled digoxin concentration, and on the right by a reciprocal plot. The rectilinear plot on the right is convenient for use with computer data processing techniques.

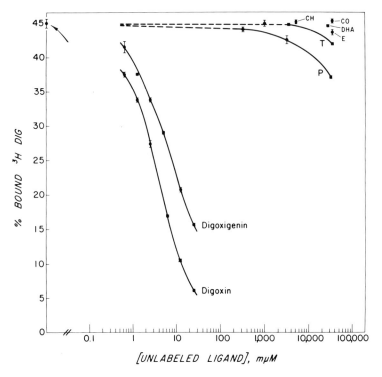

Fig. 3. Hapten inhibition data defining the specificity of a selected rabbit digoxin-specific antiserum. The extent of [³H]digoxin displacement from antibody binding sites by unlabeled digoxin, digoxigenin, cholesterol (CH), cortisol (CO), dehydroepiandrosterone (DHA), 17β-estradiol (E), progesterone (P), and testosterone (T) is shown. The endogenous steroid compounds cause measurable displacement only when present in concentrations greater than 1000-fold in excess of the amount of [³H]digoxin tracer present. [Reprinted with permission from *Biochemistry* (16).]

rected count rates for unknown samples. Concentration values for unknown samples are then printed out.

The specificity of this assay system has been documented by hapten inhibition experiments of the type shown in Figure 3. None of the endogenous steroid compounds tested produced measurable displacement of [³H]digoxin from antibody binding sites of this antiserum unless present in greater than 1000-fold molar excess. Because of cross-reactivity with certain structurally similar cardiac glycosides, however, this assay system can also be used for the determination of serum deslanoside concentrations. Known amounts of deslanoside can be used to obtain a standard curve, with tritiated digoxin serving as tracer. Radioimmunoassay techniques have also been developed for measurement of subnanogram amounts of digitoxin (18), ouabain (19), and acetyl strophanthidin (20).

TECHNICAL PROBLEMS

Several potential pitfalls appear to have been encountered by investigators who have employed radioimmunoassay techniques for the measurement of cardiac glycosides, and these bear brief discussion. Like any technique, these methods require a sound grasp of the molecular interactions involved. Meticulous attention to detail and continuous quality control are necessary to avoid erroneous results which are worse then useless to the clinician. Radioimmunoassay results are obviously no more accurate than the standards used for construction of the standard curve. Thus it is important to test crystalline digoxin (or other substance used as standard) for purity by a sensitive analytical technique such as thin-layer chromatography. The material, dried to constant weight, is then used for careful gravimetric preparation of standards. Fresh standards should be made up every few weeks if exposed to the possibility of concentration alteration by evaporation of solvent. Purity of tracer tritiated digoxin must also be carefully monitored, since there is significant variation from lot to lot from commercial suppliers.

Exogeneous radioactivity in the serum of patients, such as that introduced in radioisotope scanning procedures, must be recognized to be capable of producing erroneous results. These isotopes, usually gamma emitters, are easy to distinguish from the low-energy beta radiation emitted by tritiated tracer compounds. If one channel of the liquid scintillation spectrometer is set to pick up these higher-energy gamma events, one is immediately aware if exogenous radioactivity is present in a sample. If ^{125}I-labeled cardiac glycoside analogs are used as tracers, screening for exogenous gamma emitters in serum of patients can be achieved by directly counting an aliquot of the serum. When identified, the problem can be dealt with by extraction of the cardiac glycoside as in other procedures prior to measurement, or the method outlined by Butler (21) used to make the necessary correction.

Careful quench correction is required, since samples from clinical sources often contain bile pigments or hemoglobin among other materials that may quench significantly.

Probably the most common source of error in cardiac glycoside radioimmunoassay procedures is the use of inadequate antibody preparations. It is crucial to recognize the importance of thorough characterization of an antiserum prior to use in radioimmunoassay applications. The association constant of an antibody population is important to both as a determinant of sensitivity (22) and of stability of the antibody–hapten complex during separation of bound and free fractions. The importance of the antibody–hapten complex stability can be illustrated by a comparison of two digoxin-specific antiserums. One can determine the rate constant for dissociation of the cardiac glycoside–antibody complex by observing the rate at which a large excess of dextran-coated charcoal binds labeled glycoside tracer molecules during the dissociation phase of the equilibrium state. Figure 4 illustrates the dissociation kinetics of a high-affinity, high-specificity, digoxin-specific antiserum, designated antiserum 1, which was previously studied in detail (16). Figure 4 also shows, in contrast, the dissociation kinetics of another antiserum of inferior quality (antiserum 2), which has an association constant for digoxin more than an order of magnitude lower than that of antiserum 1. It can be seen that the complex formed by antiserum 2 dissociates rapidly compared with that formed by antiserum 1. Therefore, if charcoal is added sequentially to a series of standard and unknown samples, those in contact with charcoal for longer times will have greater dissociation of the antibody–hapten complex, hence will show fewer supernatant counts. The result will be falsely elevated estimations of serum digoxin concentration in samples exposed for longer times than the standards. This proves to be no problem with antiserum 1, but an antiserum with the dissociation kinetics of antiserum 2 requires special measures to ensure adequate control of charcoal contact time.

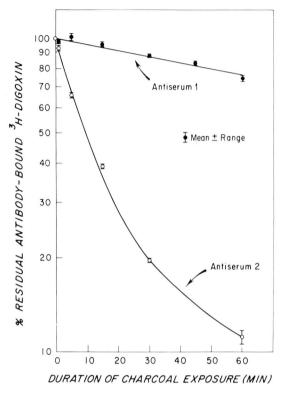

Fig. 4. Time course of dissociation of [^{3}H]digoxin–antibody complex. Percent of original binding is plotted against duration of exposure to dextran-coated charcoal. Antiserum 1, which has an average intrinsic association constant of 1.7×10^{10} liters/minute, shows slow dissociation of the complex. In contrast, antiserum 2 shows rapid dissociation kinetics which can produce significant error in radioimmunoassay use.

Figure 5 shows that despite these rapid dissociation kinetics, antiserum 2 provides a standard curve which on superficial examination appears satisfactory for analytical use. This problem of relatively rapid antibody–hapten dissociation kinetics has been shown to exist with the components supplied in a commercially available digoxin radioimmunoassay kit (23), leading Meade and Kleist to suggest an alternative means of separating bound and free digoxin. This may provide a satisfactory alternative to the use of antibodies with slower dissociation kinetics, so long as adequate specificity can be maintained.

The problem of specificity can also be illustrated by comparing the same two antiserums. Antiserum 1, as shown in Figure 3, has very high specificity for the homologous hapten digoxin. In contrast, Figure 6 shows a far lesser degree of specificity in the case of antiserum 2. Use of inferior antiserums of this type probably explains the difficulties with radioimmunoassay specificity experienced by some workers. Since commercial suppliers of radioimmunoassay materials have in general not provided the purchaser with adequate data characterizing the antiserum supplied, at this time the responsibility for testing the quality of antiserum obtained through commercial sources usually rests with the user. Minimum testing of the suitability of a given antiserum for radioimmunoassay use requires hapten in-

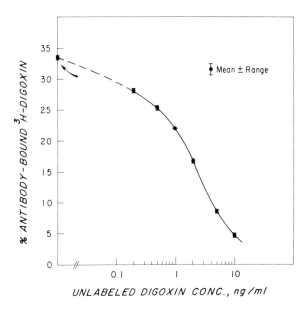

Fig. 5. Digoxin radioimmunoassay standard curve obtained using antiserum 2. This curve superficially appears adequate for radioimmunoassay use, despite the unsatisfactory characteristics of the antibody population.

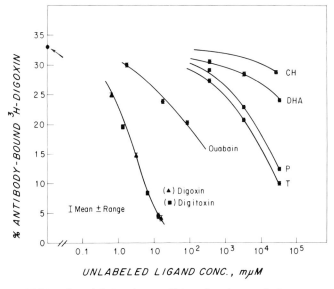

Fig. 6. Hapten inhibition data defining the specificity of antiserum 2. In contrast to antiserum 1 (Figure 3), relatively poor specificity results in displacement of [^{3}H]digoxin from antibody binding sites by relatively low concentrations of progesterone and testosterone, thus leading to potentially erroneous results if used in a radioimmunoassay system. The arrow on the ordinate denotes [^{3}H]digoxin binding in the absence of any competing ligand. CH, Cholesterol; DHA, dehydroepiandrosterone; P, progesterone; T, testosterone.

hibition studies to define specificity, and charcoal contact time studies to define the range of contact times that could be used without introducing an unacceptable degree of error. Finally, one should be aware that specificity of cardiac glycoside radioimmunoassay systems generally tends to increase with increasing duration of the incubation step (17), and this should be systematically studied in the process of setting up a radioimmunoassay procedure.

CLINICAL USE OF SERUM OR PLASMA CARDIAC GLYCOSIDE CONCENTRATION DATA

Four lines of evidence suggest that serum or plasma digitalis glycoside concentrations might be related to pharmacological or toxic effects in a clinically useful way. It is apparent that digitalis toxic rhythm disturbances, as well as most extracardiac manifestations of toxicity (24), are dose-related phenomena. Many studies are now available demonstrating that serum digitalis concentrations rise with increasing dosage (2). Therefore serum digitalis concentration and clinical state would be expected to be related, at least on a statistical basis. The studies of Doherty et al. (25) have demonstrated a relatively constant ratio between serum and myocardial digoxin concentration in animals after the attainment of serum-tissue equilibrium. This ratio was also relatively constant in humans (26). Since total myocardial digoxin content includes nonspecific as well as specific binding to receptors (27), however, total myocardial digoxin concentration should not be assumed to bear a one-to-one relationship to effect. A third element in the rationale for the clinical use of serum cardiac glycoside concentration measurements is the evidence implicating Na^+,K^+-ATPase in the mechanism of cardiac action of digitalis. Studies of the squid giant axon and the red-blood cell indicate that this plasma membrane-bound enzyme system is inhibited by cardiac glycosides only when present at the external cell surface. Caldwell and Keynes observed that sodium flux was inhibited only when ouabain was present at the outer surface of the squid giant axon (28). Analogous results were obtained by Hoffman (29) and by Perrone and Blostein (30) in experiments with human red blood cells. If the digitalis-sensitive site of Na^+,K^+-ATPase is similarly accessible from the external cell surface in the heart, it would not be surprising if it were responsive to plasma cardiac glycoside concentrations. Although a reasonably convincing case can be made for Na^+,K^+-ATPase inhibition as an important mechanism of toxic electrophysiological effects of cardiac glycosides, it should be noted that the mechanism of inotropic effect remains a subject of active controversy (31,32).

Finally, recent animal experiments have documented a close relationship between serum digoxin concentration and the electrophysiological effects of this drug on the normal canine heart. Serum digoxin concentration was found to be closely related to drug-induced changes in cardiac automaticity when tested by provocation of repetitive ventricular responses by low-energy endocardial stimulation, or by digitalis tolerance testing with infusion of the rapidly acting cardenolide acetyl strophanthidin (33).

Various laboratories, using several techniques, have reported data that are in substantial agreement concerning serum or plasma digoxin and digitoxin concentrations in patients receiving usual doses of these drugs. Table 1 shows mean serum digoxin concentrations in several reported groups of patients comprising a total of more than 1000 subjects studied. Although not shown in this table, larger doses of digoxin were associated with higher serum concentrations of the drug in all studies in which this relationship was examined (2). Renal impairment is also associated with higher serum digoxin concentrations at any given dose level (34). Some of the variation in mean serum levels is probably due to differing times at which serum was obtained for digoxin concentration determination. Mean values for nontoxic patients in these studies tend to cluster closely around the mean steady-state blood level of 1.4 ± 0.3 (S.D.) ng/ml found by Marcus et al. in a study of normal volunteers receiving oral doses of 0.5 mg of tritiated digoxin per day (35).

Mean serum or plasma digoxin concentrations in patients with electrophysiological evidence of toxicity are also listed in Table 1. Although many variables are known to influence cardiac response to digitalis glycosides, significantly higher mean serum digoxin concentrations were observed in toxic patients compared with nontoxic patients in nearly all these studies, the report of Fogelman et al.(39) being the only exception. Generally, mean digoxin concentrations in patients with toxicity are about two- to threefold higher than those of nontoxic patients. Despite statistically significant differences in mean levels, however, overlap has been observed in most series, and it should be emphasized that no single concentration can be chosen that clearly distinguishes between toxic and nontoxic serum digoxin levels. Serum concentration data are most useful when interpreted together with all other clinically relevant variables. These include serum potassium, magnesium, and calcium concentrations, adequacy of tissue oxygenation, acid-base balance, age, renal function, thyroid status, autonomic nervous system tone, other drugs concurrently

Table 1 Serum or Plasma Digoxin Concentrations: Nontoxic and Toxic Patients

Reference	Method	Mean Concentration		Statistical Significance
		Nontoxic	Toxic	
Beller et al. (1)	Radioimmunoassay	1.0	2.3	Yes
Bertler and Redfors (36)	^{86}Rb uptake	0.9	2.4	Yes
Brooker and Jelliffe (14)	Enzymic displacement	1.4	3.1	Yes
Burnett and Conklin (11)	ATPase inhibition	1.2	5.7	Yes
Chamberlain et al. (37)	Radioimmunoassay	1.4	3.1	Yes
Evered and Chapman (38)	Radioimmunoassay	1.38	3.36	Yes
Fogelman et al. (39)	Radioimmunoassay	1.4	1.7	No
Grahame-Smith and Everest (7)	^{86}Rb uptake	2.4	5.7	Yes
Hoeschen and Proveda (40)	Radioimmunoassay	0.8–1.3	2.8	Yes
Johnston et al. (41)	Radioimmunoassay	1.0	3.15	Yes
Morrison et al. (42)	Radioimmunoassay	0.76	3.35	Not stated
Oliver et al. (43)	Radioimmunoassay	1.6	3.0	Yes
Smith et al. (13)	Radioimmunoassay	1.3	3.3	Yes
Smith and Haber (34)	Radioimmunoassay	1.4	3.7	Yes
Zeegers et al. (44)	Radioimmunoassay	1.6	4.4	Yes

received, and (probably most importantly of all) type and severity of underlying cardiac disease. There is some question whether age, per se, deserves a place in this list beyond the extent to which advancing age correlates with decreasing renal function (45).

Ischemic heart disease has been particularly prevalent among patients with cardiac toxicity occurring at relatively low serum digoxin or digitoxin concentrations, in our experience (24). This is probably due, at least in part, to the electrophysiological abnormalities that accompany focal ischemia. These are in many respects similar to the electrophysiological abnormalities induced by toxic concentrations of digitalis glycosides. Clinical experience supports the concept that less digitalis is required to precipitate toxic rhythm disturbances in the diseased heart compared with the normal heart (46). However, it should be recognized that occasional patients require serum digoxin concentrations of 3 ng/ml or greater to maintain adequate slowing of the ventricular response to supraventricular tachyarrhythmias such as atrial, flutter or atrial fibrillation.

Serum digitoxin concentrations have also been studied by several groups, using a variety of assay methods. Table 2 summarizes the results of published studies of serum or plasma digitoxin concentrations. Mean serum or plasma digitoxin levels in nontoxic patients are roughly 10-fold higher than those of digoxin, largely as a result of the substantially greater serum albumin binding of digitoxin (47). Mean concentrations in toxic patients, as with digoxin, have been found to be significantly higher in studies in which statistical evaluation was carried out. Also, as in the case of digoxin, one cannot arbitrarily assign a diagnosis of digitalis intoxication to a patient on the basis of a higher than average serum digitoxin concentration, because of the many factors that influence individual response to these drugs. Patients receiving usual maintenance doses of digitalis leaf usually have serum digitoxin concentrations similar to those of patients on maintenance digitoxin, whether measured by radioimmunoassay (18) or by enzymic displacement (14).

A detailed review of the role of digitalis assay techniques in studies of the clinical

Table 2 Serum or Plasma Digitoxin Concentrations: Nontoxic and Toxic Patients

Reference	Method	Mean Concentration Nontoxic	Mean Concentration Toxic	Statistical Significance
Beller et al. (1)	Radioimmunoassay	20	34	Yes
Bentley et al. (10)	ATPase inhibition	23	59	Yes
Brooker and Jelliffe (14)	Enzymic displacement	31.8	48.8	Not stated
Lukas and Peterson (4)	Double-isotope dilution derivative	20	43–67 (range)	Not stated
Morrison and Killip (49)	Radioimmunoassay	25 (0.1 mg/day) 44 (0.2 mg/day)	53	Yes
Rasmussen et al. (50)	^{86}Rb uptake	16.6	48.7	Not stated
Ritzmann et al. (51)	^{86}Rb uptake	19	39–51 (range)	Not stated
Smith (18)	Radioimmunoassay	17	34	Yes

pharmacology of these drugs is beyond the scope of this discussion. This area has been summarized recently (2, 48), and further progress suggests that broader availability of serum and plasma cardiac glycoside concentration measurement techniques will result in an enhanced understanding of the pharmacology of these drugs in a broad spectrum of clinical settings.

REFERENCES

1. Beller, G. A., Smith, T. W., Abelmann, W. H., Haber, E., and Hood, W. B., Jr.: Digitalis intoxication: A prospective clinical study with serum levelcorrelations. *N. Engl. J. Med.* **284**:989–997, 1971.
2. Smith, T. W. and Haber, E.: The current status of cardiac glycoside assay techniques. In Yu, P. N. and Goodwin, J. F. (eds.), *Progress in Cardiology*, Vol. 11. Lea and Febiger, Philadelphia, 1973, pp. 49–73.
3. Doherty, J. E.: The clinical pharmacology of digitals glycosides: A review. *Amer. J. Med. Sci.* **255**:382–414, 1968.
4. Lukas, D. S. and Peterson, R. E.: Double isotope dilution derivative assay of digitoxin in plasma, urine, and stool of patients maintained on the drug. *J. Clin. Invest.* **45**:782–795, 1966.
5. Watson, E. and Kalman, S. M.: Assay of digoxin in plasma by gas chromatography. *J. Chromatogr.* **56**:209–218, 1971.
6. Lowenstein, J. M. and Corill, E. M.: An improved method for measuring plasma and tissue concentrations of digitalis glycosides. *J. Lab. Clin. Med.* **67**:1048–1052, 1966.
7. Grahame-Smith, D. G. and Everest, M. S.: Measurment of digoxin in plasma and its use in diagnosis of digoxin intoxication. *Brit. Med. J.* **1**:286–289, 1969.
8. Bertler, A. and Redfors, A.: An improved method of estimating digoxin in human plasma. *Clin. Pharm. Ther.* **11**:665–673, 1970.
9. Gjerdrum, K.: Determination of digitalis in blood. *Acta Med. Scand.* **187**:371–379, 1970.
10. Bentley, J. D., Burnett, G. H., Conklin, R. L., and Wasserburger, R. H.: Clinical application of serum digitoxin levels—A simplified plasma determination. *Circulation* **41**:67–75, 1970.
11. Burnett, G. H. and Conklin, R. L.: The enzymatic assay of plasma digoxin. *J. Lab. Clin. Med.* **78**:779–784, 1971.
12. Oliver, G. C., Jr., Parker, B. M., Brasfield, D. L., and Parker, C. W.: The measurement of digitoxin in human serum by radioimmunoassay. *J. Clin. Invest.* **47**:1035–1042, 1968.
13. Smith, T. W., Butler, V. P., Jr., and Haber, E.: Determination of therapeutic and toxic serum concentrations by radioimmunoassay. *N. Engl. J. Med.* **281**:1212–1216, 1969.
14. Brooker, G. and Jelliffe, R. W.: Serum cardiac glycoside assay based upon displacement of ^{3}H-ouabain from Na-K ATPase. *Circulation* **45**:20–36, 1972.
15. Butler, V. P., Jr.: Assays of digitalis in the blood. *Progr. Cardiovasc. Dis.* **14**:571, 1972.
16. Smith, T. W., Butler, V. P., Jr., and Haber, E.: Characterization of antibodies of high affinity and specificity for the digitalis glycoside digoxin. *Biochemistry* **9**:331–337, 1970.
17. Smith, T. W. and Kaplan, E.: Radioimmunoassay of cardiac glycosides: Progress and pitfalls. In Strauss, H. W., Pitt, B., and Jones, A. E., Jr. (eds.), *Nuclear Cardiology*. Johns Hopkins Symposium in Nuclear Cardiology, C. V. Mosby, St. Louis, in press, 1974.
18. Smith, T. W.: Radioimmunoassay for serum digitoxin concentration: Methodology and clinical experience. *J. Pharm. Exp. Ther.* **175**:352–360, 1970.
19. Smith, T. W.: Ouabain-specific antibodies: Immunochemical properties and reversal of Na$^+$,K$^+$-activated adenosine triphosphatase inhibition. *J. Clin. Invest.* **51**:1583–1593, 1972.
20. Selden, R., Klein, M. D., and Smith, T. W.: Plasma concentration and urinary excretion kinetics of acetyl strophanthidin. *Circulation* **47**:744–751, 1973.
21. Butler, V. P., Jr.: Digoxin radioimmunoassay. *Lancet* **1**:186, 1971.

22. Yalow, R. S. and Berson, S. A.: In Margoulies, M. *Protein and Polypeptide Hormones.* Excerpta Medica Foundation, Amsterdam, 1969, pp. 36–44, 71–76.
23. Meade, R. C. and Kleist, T. J.: Improved radioimmunoassay of digoxin and other sterol-like compounds using Somogyi precipitation. *J. Lab. Clin. Med.* **80**:748–754, 1972.
24. Smith, T. W.: Contributions of quantitative techniques to the understanding of the clinical pharmacology of digitalis. *Circulation* **46**:188–189, 1972.
25. Doherty, J. E. and Perkins, W. H.: Tissue concentration and turnover of tritiated digoxin in dogs. *Amer. J. Cardiol.* **17**:47–52, 1966.
26. Doherty, J. E., Perkins, W. H., and Flanigan, W. J.: The distribution and concentration of tritiated digoxin in human tissues. *Ann. Intern. Med.* **66**:116–124, 1967.
27. Marks, B. H.: Factors that affect the accumulation of digitalis glycosides by the heart. In Marks, B. H. and Weissler, A. (eds.), *Basic and Clinical Pharmacology of Digitalis.* Charles C Thomas, Springfield, Ill., 1972, pp. 69–93.
28. Caldwell, P. C. and Keynes, R. D.: The effect of ouabain on the efflux of sodium from a squid giant axon. *J. Physiol.* **148**:8P–9P, 1959.
29. Hoffman, J. F.: The red cell membrane and the transport of sodium and potassium. *Amer. J. Med.* **41**:666–680, 1966.
30. Perrone, J. R. and Blostein, R.: Asymmetric interaction of inside-out and right-side-out erythrocyte membrane vesicles with ouabain. *Biochim. Biophys. Acta* **291**:680–689, 1973.
31. Lee, K. S. and Klaus, W.: The subcellular basis for the mechanism of inotropic action of cardiac glycosides. *Pharm. Rev.* **23**:193–261, 1971.
32. Okita, G. T., Richardson, F., and Roth-Schecter, B. F.: Dissociation of the positive inotropic action of digitalis from inhibition of sodium- and potassium-activated adenosine triphosphatase. *J. Pharmacol. Exp. Ther.* **185**:1–11, 1973.
33. Barr, I., Smith, T. W., Klein, M. D., Hagemeijer, F., and Lown, B.: Correlation of the electrophysiologic action of digoxin with serum digoxin concentration. *J. Pharmacol. Exp. Ther.* **180**:710–722, 1972.
34. Smith, T. W. and Haber, E.: Digoxin intoxication: The relationship of clinical presentation to serum digoxin concentration. *J. Clin. Invest.* **49**:2377–2386, 1970.
35. Marcus, F. I., Burkhalter, L., Cuccia, C., Pavlovich, J., and Kapadia, G. G.: Administration of tritiated digoxin with and without a loading dose: A metabolic study. *Circulation* **34**:865–874, 1966.
36. Bertler, A. and Redfors, A.: Plasma levels of digoxin in relation to toxicity. *Acta Pharm. Tox.* **29**(suppl. III):281–287, 1971.
37. Chamberlain, D. A., White, R. J., Howard, M. R., and Smith, T. W.: Plasma digoxin concentrations in patients with atrial fibrillation. *Brit. Med. J.* **3**:429–432, 1970.
38. Evered, D. C. and Chapman, C.: Plasma digoxin concentrations and digoxin toxicity in hospital patients. *Brit. Heart J.* **33**:540–545, 1971.
39. Fogelman, A. M., LaMont, J. T., Finkelstein, S., Rado, E., and Pearce, M. L.: Fallibility of plasma-digoxin in differentiating toxic from non-toxic patients. *Lancet* **2**:727–729, 1971.
40. Hoeschen, R. J. and Proveda, V.: Serum digoxin by radioimmunoassay. *Can. Med. Assoc. J.* **105**:170–173, 1971.
41. Johnston, C. I., Pinkus, N. B., and Down, M.: Plasma digoxin levels in digitalized and toxic patients. *Med. J. Aust.* **1**:863–866, 1972.
42. Morrison, J., Killip, T., and Stason, W. B.: Serum digoxin levels in patients undergoing cardiopulmonary bypass. *Circulation* **42**:III, 110, 1970.
43. Oliver, G. C., Parker, B. M., and Parker, C. W.: Radioimmunoassay for digoxin. Technic and clinical application. *Amer. J. Med.* **51**:186–192, 1971.
44. Zeegers, J. J. W., Maas, J. H. J., Willebrands, A. F., Kruyswijk, H. H., and Jambroes, G.: The radioimmunoassay of digoxin. *Clin. Chim. Acta* **44**:109–117, 1973.
45. Ewy, G. A., Kapadia, G. G., Yao, L., Lullin, M., and Marcus, F. I.: Digoxin metabolism in the elderly. *Circulation* **39**:449–453, 1969.
46. Smith, T. W. and Willerson, J. T.: Suicidal and accidental digoxin ingestion: Report of five cases with serum digoxin level correlations. *Circulation* **44**:29–36, 1971.

REFERENCES

47. Lukas, D. S. and DeMartino, A. G.: Binding of digitoxin and some related cardenolides to human plasma proteins. *J. Clin. Invest.* **48**:1041–1053, 1969.
48. Butler, V. P., Jr.: Assays of digitalis in the blood. *Progr. Cardiovasc. Dis.* **14**:571, 1972.
49. Morrison, J. and Killip, T.: Radioimmunoassay of digitoxin. *Clin. Res.* **14**:668, 1970.
50. Rasmussen, K., Jervell, J., and Storstein, O.: Clinical use of a bio-assay of serum digitoxin activity. *Eur. J. Clin. Pharm.* **3**:236–242, 1971.
51. Ritzmann, L. W., Bangs, C. C. Coiner, D., Custis, J. M. and Walsh, J. R.: Serum glycoside levels in digitalis toxicity. *Circulation* **40**:III, 170, 1969.

Renin Activity Assay: Angiotensin I Generation and Radioimmunoassay

CHAPTER FIVE JEAN E. SEALEY, B.Sc.
JOHN H. LARAGH, M.D.

Prior to the last few years, plasma renin activity was measured by an indirect method which utilized an incubation procedure during which angiotensin was generated by the action of renin on plasma renin substrate. The angiotensin formed during incubation was then assayed using a rat pressor bioassay (1).

More recently, with the development of the radioimmunoassay technique by Yalow and Berson (2), the rat pressor bioassay has been replaced by radioimmunoassay of the formed angiotensin (1,3,4–6). However, we still do not have a direct assay for renin, and it is still quantitated by utilizing the fact that as an enzyme it reacts with plasma renin substrate to form angiotensin I (Figure 1). It is the angiotensin I generated during an incubation step that is quantitated by radioimmunoassay.

The change from bioassay to radioimmunoassay has introduced a new series of problems. When the bioassay technique was used, it was not important whether angiotensin I or angiotensin II was the end product of the incubation step, since angiotensin I is converted to angiotensin II by the rat's own converting enzyme. However, the radioimmunoassay recognizes either angiotensin I or angiotensin II, but not both, and the conversion of angiotensin I to angiotensin II must be inhibited during the incubation step. Angiotensin I was chosen for detection in the radioimmunoassay rather than angiotensin II because converting enzyme in blood is rate limiting, and quantitative conversion of angiotensin I to II is not possible in unextracted plasma.

In addition, radioimmunoassay has introduced another problem that was not present in the bioassay era. Antibodies to angiotensin I recognize nonspecific substances in plasma. A blank must be subtracted from each value unless sufficient angiotensin I is generated during the incubation to overcome the blank. This blank subtraction leads to considerable error in the assay of samples with low plasma renin activity.

Over the years we have developed a renin incubation procedure which is very

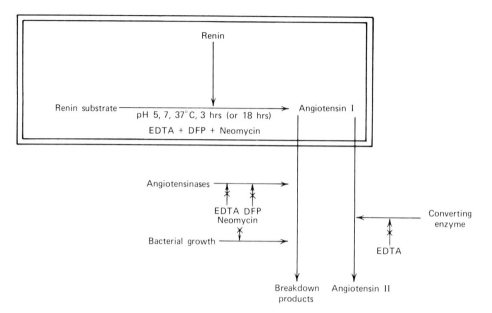

Fig. 1. Plasma renin activity measurement: incubation for generation of angiotensin I. This figure outlines the various steps involved in generation and destruction of angiotensin I, and suggests protective agents that can be used to prevent destruction of formed angiotensin.

sensitive and accurate, so that maximum amounts of angiotensin can be generated (1,3,7). This in turn reduces the problems and increases the sensitivity of the immunoassay. We first consider optimum conditions for collection of blood for renin assay, then go into details of our renin incubation and our immunoassay, and finally consider the problems now posed by commercially available kits.

NORMALCY OF PLASMA RENIN MEASUREMENTS

The normal range of plasma renin activity is defined relative to the concurrent 24 hour urinary sodium excretion (Figure 2) (8,9). This nomogram was derived from the study of 52 normal volunteers. Urinary sodium excretion is taken as an index of sodium balance. It is apparent that during sodium loading (urinary sodium excretion > 150 mEq/24 hours) normal renin values can be very low, and they sometimes approach the limits of detectability. Sodium depletion, however, stimulates renin release, so that the relationship of plasma renin activity to the concurrent 24 hour urinary sodium excretion takes the form of a hyperbolic curve. Because of the close interdependence of plasma renin levels and sodium balance, definition of normality and classification of hypertensive patients into subgroups according to their plasma renin activity is meaningful only if an index of sodium balance, for example, urinary sodium excretion rate, is taken into account (8,9).

The physiological circumstances under which renin is measured must be controlled. Ideally, the patient should be off all antihypertensive and diuretic drugs for several weeks, and should ingest a moderately low amount of sodium. Also, since

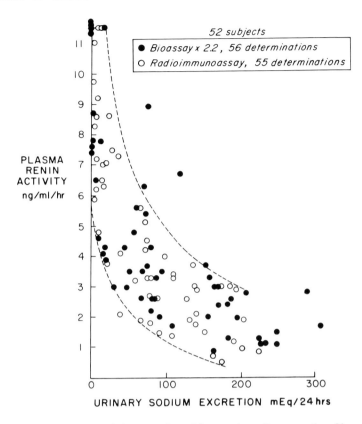

Fig. 2. The normal relationship of plasma renin activity to urine sodium excretion. Since plasma renin activity varies according to the state of sodium balance, normalcy of values can be assessed only in relationship to some indicator of the state of sodium balance. The 24 hour urine sodium excretion is a simple and useful indicator of sodium balance, and a complete 24 hour urine collection should accompany each plasma renin activity measurement.

other reported methods for plasma renin may give considerably lower values, when these methods are used even greater sodium restriction may be required to detect the low values (3).

For patients with low-renin essential hypertension, the renin value is the only consistent feature that distinguishes them from patients with other forms of essential hypertension. Methods of the greatest sensitivity are therefore required so that truly low values can be discriminated from those falling in the normal range.

COLLECTION OF BLOOD

Blood is collected in 20 ml EDTA Vacutainers containing 0.17 ml of 15% potassium EDTA from subjects in the upright position for 4 hours. The tubes are inverted and chilled immediately. The blood or plasma is never allowed to come to room temperature after this time. After centrifugation at 4°C for 20 minutes, the plasma is separated and stored at $-20°C$ until incubation for generation of angiotensin I.

Table 1 Incubation for Generation of Angiotensin I

2 ml plasma containing 0.003 M EDTA
0.04 ml 10% neomycin sulfate
2 drops DFP (1/20 dilution in isopropyl alcohol)[a]
1, 0.5, or 0.1 N HCl to pH 5.7
3 or 18 hours incubation at 37°C
Freeze
Radioimmunoassay

[a] Add using a 4 inch, 19 gauge needle attached to a 1 ml disposable syringe.

RENIN INCUBATION PROCEDURE

The incubation for generation of angiotensin I is carried out as described in Figure 1 and Table 1. EDTA, added during collection of plasma, serves as anticoagulant, and its concentration (0.003 M) is sufficient to inhibit converting enzymes, and to some extent angiotensinases, during the incubation step. If the Vacutainer is not completely filled with blood, the resultant higher concentration of EDTA does not affect the assay. Potassium rather than sodium EDTA is chosen because it is present in the Vacutainer as a 15% solution. Powdered sodium EDTA used in other Vacutainers is rather insoluble and might result in an insufficient concentration of EDTA in the plasma.

0.04 ml of 20% DFP is added to 2 ml of plasma to inhibit other angiotensinases, and 40 μl of 10% neomycin sulfate is added to retard bacterial growth. The pH of each sample is individually adjusted to 5.7 with hydrochloric acid. For samples with low plasma renin activity the plasma is incubated in a shaker-type water bath for 18 hours at 37°C, rather than for the 3 hours used routinely.

VALIDATION OF USE OF pH 5.7

The pH optimum for human renin with plasma renin substrate was found to be between pH 5.5 and 6.5 (Figure 3), confirming previous reports (10). The rate of angiotensin generation fell steeply below pH 5.5 and declined less steeply above pH 6.5. At pH 7.4 the rate of angiotensin generation was almost half that found at the pH optimum. In addition to a lower rate of production of angiotensin I, the pH of the plasma did not remain stable when the incubation was carried out above pH 6.0, and in some samples it drifted high enough to almost stop the action of renin. This increase in pH during incubation may be due to slow elimination of carbon dioxide from plasma. The lack of change in pH at more acid pH may have been because carbon dioxide was eliminated from the plasma by the addition of acid prior to incubation.

We recommend that the incubation step for generation of angiotensin I be carried out at the pH optimum, so that the maximum rate of generation of angiotensin is

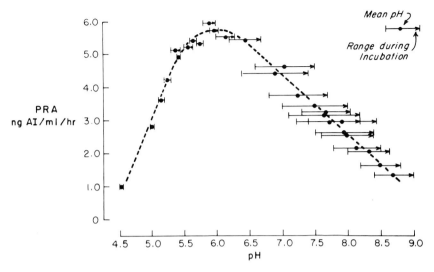

Fig. 3. pH optimum of human renin with human renin substrate. All samples were incubated for 3 hours with EDTA, DFP, and neomycin in a 37°C water bath which was shaken throughout the incubation. The pH optimum is quite flat between 5.5 and 6.5. Arrows represent the change in pH that occurred during the incubation. The pH remains quite stable below 6.0, but can increase to such an extent that during incubation at alkaline pH the action of renin can stop.

achieved. The pH should be adjusted to below pH 6.0, so that if perchance the pH does change slightly during the incubation, the rate of generation of angiotensin will not vary since the flat portion of the pH optimum curve has been exploited. Alkaline pH should be avoided because of the lower rate of angiotensin generation, and because instability of pH during incubation considerably alters the rate of angiotensin generation. The incubation should never be carried out on unadjusted plasma, since the pH of plasma collected in EDTA can range from 7.4 to 8.5, and differences in the rate of angiotensin generation may be entirely due to differences in the pH of incubation and not to differences in plasma renin activity.

LINEARITY OF AN 18-HOUR INCUBATION

At pH 5.7 in the presence of EDTA, DFP, and neomycin, the rate of angiotensin generation is linear for up to 18 hours (Figure 4). In Figure 4 the hourly rate of angiotensin generation for samples incubated for 3 hours is entirely similar to that found in the same samples when incubated for 18 hours. Thus complete angiotensinase and converting enzyme inhibition is achieved for up to 18 hours.

The addition of a bacteriostatic agent to the incubation medium is important for prolonged incubations. Shaking the samples during incubation also promotes linearity in the rate of angiotensin generation possibly because of dispersion of the slight precipitate that forms during incubation.

Since the rate of angiotensin generation is linear for up to 18 hours, all samples that have plasma renin activity less than 1.0 ng/ml per hour are reincubated for 18 hours. In this way the amount of angiotensin generated in low-renin samples is

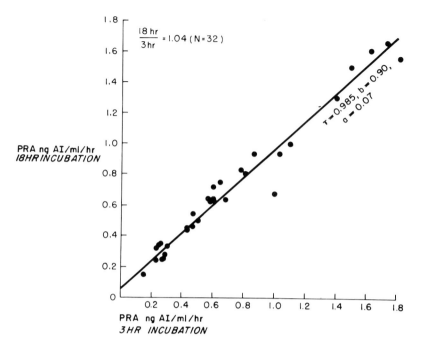

Fig. 4. Comparison of the rate of angiotensin generation in the same samples incubated for 3 or 18 hours under the same conditions. The rate is similar for both time periods, thus justifying the use of an 18 hour incubation period for samples with low plasma renin activity.

similar to that generated in normal renin samples incubated for 3 hours. Thus low and normal renin samples can be detected with equal accuracy. The use of an 18 hour incubation is technically very easy, since the samples can be left unattended to incubate overnight.

RADIOIMMUNOASSAY FOR QUANTITATION OF ANGIOTENSIN I

A 0.1 M concentration of tris buffer (pH 7.5) is used throughout the radioimmunoassay procedure (1,7). Protein is added to the buffer to inhibit adsorption of angiotensin onto glassware and plastic containers, and to increase the stability of the diluted antibody. Considerable adsorption to plastic has been observed when the protein concentration is reduced below 0.4%. Neomycin and phenylmecuric acetate are added to retard bacterial growth.

A unique feature of this particular radioimmunoassay is the prior mixing of radioactive angiotensin with antibody. This step simplifies the radioimmunoassay considerably, and greatly reduces inaccuracies caused by multiple pipettings. Radioactive angiotensin containing approximately 250,000 cpm is added to 200 ml of tris buffer and mixed thoroughly. Ten milliliters of mixture is removed prior to addition of antibody for measurement of nonspecific binding of angiotensin to proteins in the buffer solution. Antibody to angiotensin I should be added to the mixture less than an hour before use, because prolonged storage may lead to reduced sensitivity.

Prior to radioimmunoassay the samples may be diluted in tris buffer if the renin value is expected to be above the normal range. However, for most samples no predilution is necessary.

Since angiotensin I loses activity when stored in dilute solution, the standards used in the radioimmunoassay are prepared fresh daily from a stock containing 10 μg/ml. The stock solution and the diluted standards are prepared in 0.1 M tris buffer containing 0.6% protein and bacteriostatic agents. Each month the 10 μg/ml stock solution is prepared from a 50 μg/ml solution which was weighed out and diluted in boiled distilled water containing 0.2% neomycin sulfate.

Ten tubes in duplicate are included in the standard curve (Table 2). These include seven pairs of tubes with different amounts of angiotensin I standard, as well as tubes in duplicate for measuring the total counts, the nonspecific binding of angiotensin to proteins in the buffer solution, and the percent binding of radioactive angiotensin when no added cold angiotensin I is present (B_0).

At least 20 other tubes are prepared in duplicate for each plasma sample to be assayed and for two standard plasma samples. The latter are included for quality control purposes. Incubated plasma samples of 10 and 20 μl each are assayed.

Two milliliters of solution without antibody is added to tubes 17 and 18 (Table 2)

Table 2 Production of the Standard Curve[a]

Tube no.	Angiotensin Standard (ng/ml)	Volume Added (μl)	Angiotensin (pg)
1, 2	1	10	10
3, 4	1	20	20
5, 6	4	10	40
7, 8	4	20	80
9, 10	15	10	150
11, 12	15	20	300
13, 14	20	20	400
15, 16	—	Total counts; no charcoal;	
17, 18	—	nonspecific binding; no antibody (B_0); no added angiotensin	
19, 20	—		
Unknown samples			
101	—	10	—
101'	—	20	—
102	—	10	—
102'	—	20	—

[a] Addition of reagents: 2 ml radioactive angiotensin–antibody mixture added to each tube, except nos. 17 and 18. Separation of bound from free: After 18 hours at 4°C, 0.5 ml of charcoal mixture is added to each tube except nos. 15 and 16.

for measurement of nonspecific binding. Two milliliters of the radioactive angiotensin–antibody mixture is added to all other tubes except nos. 17 and 18. A Micromedic diluter may be used for the addition of this mixture to the standards and unknown samples. The samples are mixed gently on a vortex mixer if the automatic diluter is not used. The tubes are covered with Parafilm and then placed in a refrigerator at 4°C for 18 hours.

After 18 hours for equilibration of angiotensin with antibody, the angiotensin bound to antibody is separated from the free by adsorption of the free angiotensin onto dextran-coated charcoal. Because of the tendency of charcoal to slowly absorb a small amount of bound angiotensin, the duplicates of the standard curve are divided so that charcoal is added to one half at the beginning and to the other half at the end of the series. After centrifugation the supernatant from each tube is counted for 10 minutes each.

CALCULATIONS

The relationship of bound to free angiotensin I is calculated using a modified logit plot (Figure 5). The counts bound in the standard tubes are divided by the mean counts bound in the tubes 19 and 20 to which no angiotensin was added (B/B_0). $(B/B_0)/(1 - B/B_0)$ is then calculated and plotted against the amount of angiotensin I in each tube of the standard curve. When log/log paper is used, a straight-line relationship is achieved.

The amount of angiotensin in unknown samples is then derived from the standard curve, and then plasma renin activity is calculated taking into account (*a*) the volume of plasma assayed (10 or 20 µl); (*b*) the dilution of plasma due to addition of DFP, neomycin, and acid to adjust the pH, (*c*) dilution of incubated plasma prior to radioimmunoassay, if any, and (*d*) incubation time (3 or 18 hours). Plasma renin activity is expressed as nanograms of angiotensin I generated per milliliter per hour.

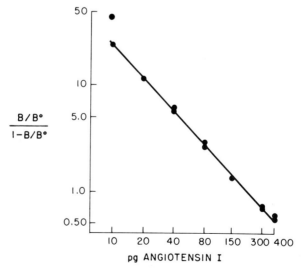

Fig. 5. Modified logit plot of angiotensin I standard curve. Values are plotted on log/log paper.

TROUBLE SHOOTING

Radioimmunoassay involves many steps, and when variability occurs it can be ascribed to many different problems. A useful guide to changes in the radioimmunoassay is nonspecific binding (Table 2, tubes 17 and 18). If the duplicate counts of these tubes are variable, this may be due to erratic addition of charcoal. The addition of charcoal for separation of bound from free is the source of many errors in this particular radioimmunoassay. The charcoal should be stirred vigorously, and every effort should be made to add it as quickly and consistently as possible.

If the nonspecific binding gradually and consistently increases with each set—it should be 2–3% of the total counts—this is a sign that the radioactive angiotensin is deteriorating, and fresh label is recommended. When changing to a new batch of charcoal, the nonspecific binding may increase or fall. Each batch of charcoal is different in activity, and the optimum amount should be calculated for each batch. If too much is added, the adsorption of antibody-bound angiotensin will increase to unacceptable levels, and if too little is used erratic results will be achieved.

Another guide to the adequacy of the label is the variability of the standards at the extremes of the standard curve. As the label ages, the sensitivity of the standard curve decreases, and the lowest point often fails to fall on the straight line of the modified logit plot. Also, if the amount of antibody added is too much or too little, the points at the extremes of the standard curve will not fall on the straight line. Binding of 50% is the optimum amount. Only the portion of the standard curve that falls on the straight line is used in the assay. Samples that assay outside this range should be repeated.

Because angiotensin does not retain its activity when stored in dilute solution, the standards used in the radioimmunoassay are prepared fresh daily. This can lead to error, and it is especially important because of this that two standard plasma samples be run with each set. If these do not fall within a predetermined range, the set should be discarded and repeated.

ASSAY SENSITIVITY

The adequacy of the angiotensin I assay is determined by the sensitivity of the radioimmunoassay for detecting angiotensin I, and also by the specificity of the antibody utilized in the radioimmunoassay. Unlike many other radioimmunoassay procedures, such as the angiotensin II assay, sensitivity of the radioimmunoassay system need not be a problem in the case of renin. That is because the amount of angiotensin generated can be increased until there is sufficient to be measured accurately (3). Very few methods take advantage of this great potential. Thus even though our radioimmunoassay step has a sensitivity similar to other methods, we can detect much lower levels by prolonging the time of incubation to 18 hours and optimizing the conditions of the incubation step.

The ability to prolong the incubation time eliminates a problem inherent in radioimmunoassay. Most antibodies to angiotensin I cross-react with endogenous angiotensin I or with other nonspecific substances in plasma. Renin activity is defined by the rate of angiotensin generation during a controlled incubation period. If angiotensin or other immunoreactive substances are present in unincubated plasma, the renin activity value will be falsely high unless the blank is subtracted. By prolonging the incubation time it is possible to ignore the blank, because the amount of angiotensin generated during incubation is far in excess of the blank and completely masks it.

COMMERCIAL KITS

If commerical kits are used as a source of reagents for the immunoassay, the following steps should be incorporated into the procedure. (1) The incubation step should be changed to pH 5.7 in the presence of DFP and neomycin, as well as EDTA. (2) For samples with low plasma renin activity the incubation step should be repeated for 18 hours. (3) Minimum dilution of plasma should be attempted during pH adjustment. (4) Sufficient angiotensin I should be generated to eliminate the necessity for blank subtraction. (5) A minimum volume of plasma should be used in the radioimmunoassay. (6) A large source of antibody should be acquired, if possible, so that variability in antibody characteristics does not become a problem in assaying low-renin samples. (7) For routine laboratories that have little control over the temperature of the blood after collection (it should be chilled) assay of the blank should not be discontinued, since high blank values expose samples that have been left unchilled for an extensive period of time. (8) Strict quality control is essential in any radioimmunoassay, so that the same two plasma samples should be run with each set and, if they vary by more than $\pm 15\%$, the set should be discarded and then repeated.

SUMMARY

A method for measurement of plasma renin activity is described in which angiotensin I, generated during a 3 or 18 hour incubation period is quantitated by radioimmunoassay. The incubation step is carried out in undiluted plasma at the pH optimum (pH 5.7) with complete angiotensinase and converting enzyme inhibition by EDTA and DFP. Blank subtraction is eliminated by the generation of large amounts of angiotensin during incubation, which also increases the accuracy of the measurements.

Recommendations for modification of commercially available kits include changes in the incubation step for generation of angiotensin I. A trouble-shooting section is included in which changes in the radioimmunoassay step are pointed out, which provide clues to possible errors and to sources of variability.

REFERENCES

1. Sealey, J. E., Gerten-Banes, J., and Laragh, J. H.: The renin system: Variations in man measured by radioimmunoassay or bioassay. *Kidney Int.* **1**:240–253, 1972.
2. Yalow, R. S. and Berson, S. A. Immunoassay of endogenous plasma insulin in man. *J. Clin. Invest.* **39**:1157, 1960.
3. Sealey, J. E. and Laragh, J. H. Searching out low renin patients: Limitations of some commonly used methods. *Amer. J. Med.* **55**:303–314, 1973.

REFERENCES

4. Boyd, G. W., Adamson, A. R., Fitz, A. E., and Peart, W. S. Radioimmunoassay determination of plasma-renin activity. *Lancet* **1**:213–218, 1969.
5. Haber, E., Koerner, T., Page, L. B., Kliman, B., and Purnode, A. Application of a radioimmunoassay for angiotensin I to the physiologic measurements of plasma renin activity in normal human subjects. *J. Clin. Endocrinol.* **29**:1349–1355, 1969.
6. Cohen, E. L., Grim, C. E., Conn, J. W., Blough, W. M., Jr., Guyer, R. B., Kem, D. C., and Lucas, C. P. Accurate and rapid measurement of plasma renin activity by radioimmunoassay. *J. Lab. Clin. Med.* **77**:1025–1038, 1971.
7. Sealey, J. E., Laragh, J. H., Gerten-Banes, J., and Aceto, R. M. The measurement of plasma renin activity in man. In J. H. Laragh (ed.), *Hypertension Manual.* Yorke Publishing Company, New York, 1974, pp. 621–640.
8. Brunner, H. R., Laragh, J. H., Baer, L., Newton, M. A., Goodwin, F. T., Krakoff, L. R., Bard, R. H., and Bühler, F. R. Essential hypertension: Renin and aldosterone, heart attack and stroke. *N. Engl. J. Med.* **286**:441–449, 1972.
9. Laragh, J. H., Baer, L., Brunner, H. R. Bühler, F. R., Sealey, J. E., and Vaughan, E. D., Jr. Renin, angiotensin and aldosterone system in pathogenesis and management of hypertensive vascular disease. *Amer. J. Med.* **52**:633–652, 1972.
10. Pickens, P. T., Bumpus, F. M., Lloyd, A. M., Smeby, R. R., Page, I. H. Measurement of renin activity in human plasma. *Circ. Res.* **17**:438–448, 1965.

The Detection of Hepatitis B Antigen (HBAg) by Radioimmunoassay

CHAPTER SIX   **WILLIAM H. BRINER, Captain, USPHS (Retired)**

Hepatitis B antigen (HBAg, HAA, Australia antigen, hepatitis-associated antigen, SH antigen) was first detected in the blood of an Australian aborigine by Blumberg and associates in 1963 (1). Their subsequent finding of a correlation between this antigenic substance and long-incubation hepatitis (2) has stimulated much additional research investigation into the etiology of viral hepatitis, both in the United States and abroad. Although the theory was held for many years that long-incubation hepatitis (serum hepatitis) was transmitted primarily by the administration of contaminated blood, or by other forms of parenteral therapy, it is now recognized that the disease is also readily transmissable by a nonparenteral route (3). The fecal-oral route has been implicated as an important means of infection (4, 5), and indeed this route may be much more important than was previously recognized.

It is clear that the transfusion of blood constitutes a potentially serious threat with regard to the transmission of long-incubation hepatitis, and the fact that 9,349,000 units of whole blood were collected in 1971 (6) merely serves to emphasize this point. Considerable attention has been directed toward improving the methods of identifying this antigenic substance in donor blood to prevent its administration to patients.

Studies have shown that a higher incidence of HBAg is found in the blood of paid or commercial donors (7,8), as contrasted with blood obtained from volunteer donors. A study conducted for the National Heart and Lung Institute indicated that, although only 17,000 cases of overt posttransfusion hepatitis were reported during the year included in the study (resulting in about 850 deaths), up to 100,000 cases of subclinical heptatitis infection due to transfusions may have occurred. The study also showed that, although less than 15% of the total supply of whole blood and blood components comes from commerical sources, it accounts for 25–45% of the reported cases of posttransfusion hepatitis. Thus it is the intent of the federal government to revise labeling regulations for blood and blood derivatives to indicate whether the donor was a voluntary one or a commercial source, and to bring about an all-volunteer blood plan as rapidly as possible (9).

HEPATITIS B ANTIGEN (HBAg)

DETECTION METHODS FOR HBAg

Various immunological test methods have been employed in the detection of HBAg. These include:

1. Immunodiffusion (ID) or agar gel diffusion (AGD).
2. Counterelectrophoresis (CEP).
3. Complement fixation (CF).
4. Hemagglutination inhibition (HAI).
5. Hemagglutination (HA).
6. Radioimmunoassay (RIA).
    a. Indirect.
    b. Direct.

If a test to indicate the presence of HBAg is to be useful, it should be sensitive, reliable, convenient (amenable to some degree of automation), rapid, and inexpensive. Although "zero defect" has become a very popular term in our society, it is conceptually very difficult to attain. Nevertheless, the direct radioimmunoassay of HBAg approaches this concept much more closely than other methods cited.

Certainly, the direct RIA method is sensitive, for its sensitivity approximates 1 ng (10). It is also inexpensive, or certainly no more costly than other test modalities—particularly when performed in nuclear medicine laboratories where the required nuclear radiation detection equipment is already present. Automation is readily available to laboratories that perform sufficient numbers of tests to make this a requirement.

Although the reliability and sensitivity of the direct RIA method for the detection of HBAg has been cited in various publications (11–14), questions have been raised concerning the ability of this test to diminish posttransfusion hepatitis (15). For the most part, these questions relate to the number of false positive tests noted in the early studies completed with this agent. A number of these seemed related to the presence of an antibody to guinea pig protein.

The direct RIA method has been criticized by blood bank personnel because of the rather lengthy period of time necessary to complete the commercially available test (16). When first licensed, this test required approximately 20 hours for completion. However, the Bureau of Biologics, U.S. Food and Drug Administration has recently approved the reduction in incubation times involved in the test, so that completion now requires only 3–4 hours. This has been accomplished with no reduction in sensitivity; indeed, sensitivity has increased approximately twofold.

THE SOLID-PHASE DIRECT RIA METHOD

This procedure uses a "sandwich" principle, consisting of two incubation steps. First, patient serum is added to a tube coated with hepatitis-associated antibody (HBAb, anti-HBAg), and incubated for an appropriate period of time at a specified temperature. During this incubation, HBAg that may be present in the serum complexes with the antibody on the tube. At the conclusion of the incubation period, the tube is rinsed with a tris buffer and aspirated, leaving only the antigen which is bound to the antibody on the tube. The "sandwich" is then completed with the addition of HBAb labeled with ^{125}I, followed by a second incubation. During this step, the labeled antibody links with available binding sites on the previously bound antigen. Thus the "sandwich" is constituted in this manner:

$$Ab_t\text{---}HBAg\text{---}HBAb$$

where Abt is the antibody coated on the tube, HBAg is the antigen in the patient serum and HBAb is the ^{125}I-labeled antibody. After the second incubation, the tube is again rinsed with tris buffer, aspirated, and counted in a well scintillation counting system. Interpretation of the test involves a comparison of counts obtained on patient serum samples with those obtained on serum known to be negative with respect to HBAg.

There are now two procedures that have been approved by regulatory authorities for use in commercially available kits (16). The first (and preferable) one stipulates an incubation temperature of 45°C for 2 hours for the first incubation, and 45°C for 1 hour for the second incubation. This procedure results in a more rapid and more sensitive test than the alternate procedure which employs ambient temperature (approximately 25°C) and incubation times of 16 hours and 90 minutes for the first and second incubations, respectively. Although plasma samples may be used in the latter (longer) procedure, serum is required for the procedure employing a shorter incubation at elevated temperature. The use of plasma at the higher temperature results in clot formation, which may trap radioactivity and thus produce false positive results.

Any repeatable positive test should be confirmed by a neutralization technique to indicate specificity prior to informing any donor or patient that he is a carrier of HBAg. The confirmatory test involves the neutralization of the previously reactive serum sample with HBAb. A confirmatory neutralization test kit containing HBAb is available from manufacturers of HBAg kits. Thus the specific inhibition of the antigen present in the patient serum by the added antibody from the confirmatory neutralization kit identifies the serum as being HBAg-positive.

RELATIVE SENSITIVITY OF THE SOLID-PHASE DIRECT RADIO-IMMUNOASSAY AND OTHER METHODS

Individual studies intended to indicate the relative sensitvity of the several available methods for the detection of HBAg may yield results that are misleading at best, particularly if one attempts to correlate individual laboratory results with those obtained in a different geographical area. It is well known that a wide variation exists in sera obtained from different areas of the world, with respect to subtypes, concentrations, and mixtures of antigens and antibodies. However, an approximation of relative sensivity may be obtained from a single study, if one acknowledges that the data so derived apply only to the specific set of conditions that prevailed during that test. Ling and Overby (10) studied the sensitivity of several test procedures in serial dilutions of a serum sample known to be positive with respect to HBAg. Their results indicated that the solid-phase direct RIA method is

125 times more sensitive then a complement fixation method, 250 times more sensitive than counter electrophoresis, and 500 times more sensitive than agar gel diffusion. Other investigators have demonstrated an increased sensivity of the RIA method in other areas and under different test conditions.

OTHER USES

Although the use of radioimmunoassay in HBAg detection has received the greatest attention in the screening of potential blood donors, there are other clinical and environmental situations in which the test has merit. It is well known, for example, that hepatitis is one of the more important laboratory-acquired infections among clinical laboratory and blood bank personnel (17,18). Thus it is prudent to maintain a regular monitoring program on these personnel to detect possible infection.

In addition, there is now a well-established prevalence of hepatitis in the patients and staff of renal dialysis units. In a report issued by the U.S. Center for Disease Control, 80% of the dialysis units surveyed over a 4 year period indicated the presence of hepatitis in patients or staff (19). Indeed, Marmion (20) reports that hepatitis ranks third in the infective causes of death in patients on regular dialysis, and Blumberg (21) has found that HBAg may appear in the blood before any other symptoms or signs of hepatitis develop in dialysis patients. Thus it is frequently possible to diagnose the disease before patients become clinically ill.

HBAg is also known to be carried by significant numbers of drug addicts (22), as well as others habituated to nonnarcotic drugs (23). A surveillance program instituted in treatment clinics can be exceedingly helpful in the management of these patients.

Food handlers who are HBAg carriers, as well as dentists and surgeons who are HBAg-positive, are potentially capable of transmitting hepatitis to people with whom they come in contact (24). Therefore the testing of these groups may also be fruitful. Unquestionably, as more is learned about the etiology and transmission of hepatitis, other population groups who may require routine examinations for subclinical disease or the carrier state may be identified.

SAFETY PRECAUTIONS

Since there now seems little doubt regarding the association of infectivity with HBAg, there is a well-defined need for a safety protocol in any laboratory where blood and patient sera are handled. The possibility of infection through direct contact with a hepatitic patient has been acknowledged in hospitals for many years. Also, the ergasteric (18) and iatrogenic (24) modes of transmission of the disease have been reported. However, an undetected HBAg carrier poses the greatest threat to hospital personnel, particularly if routine hospital procedures are inadequate to prevent the spread of disease (25). The problem becomes even more acute if one is working with patient blood and sera suspected of being HBAg-positive.

Disinfection and sterilization procedures are problematical to say the least, because of the resistance of this infectious agent. Heat sterilization is the only procedure that has been routinely effective (26). Recommendations are offered (25) that are based primarily on results obtained with other viruses:

1. Use disposable supplies wherever possible when contact with blood occurs.
2. Nondisposable medical apparatus should be sterilized between uses; autoclaving is preferable to ethylene oxide, although either is preferable to liquid sterilization.
3. If liquid sterilizing agents must be used, aqueous 40% formalin, activated glutaraldehyde, or hypochlorites are probably the agents of choice—prolonged contact with the active agent is necessary.

The protection of laboratory personnel is also of extreme importance. The following suggestions have been postulated by experts (24,25).

1. Personnel having contact with blood products should wear protective clothing, such as laboratory coats for laboratory and venipuncture personnel.
2. Personnel having direct contact with specimens known to be HBAg-positive or with blood from high-risk sources should wear protective gloves; if aerosolization of the blood is likely during a laboratory procedure, masks and gowns may offer additional protection.
3. Personnel with cuts or abrasions below the forearm should wear gloves at the time of any potential contact with blood.
4. Personnel should wash hands vigorously after any direct contact with blood or blood products.
5. Eating, drinking, and smoking should be prohibited in areas where blood contamination occurs.
6. Protective clothing should be removed before eating, drinking, or smoking.
7. Personnel should exert vigorous efforts to avoid contamination of themselves or the laboratory environment because of the potentially difficult task of decontamination and the risk of infection.
8. All accidents, injuries, and potentially contaminating incidents should be reported to a supervisor immediately.

Safety of personnel involved in the handling of blood and blood products must always be a subject of constant attention. Nuclear medicine personnel will surely note the many similarities in microbiological and radiological safety concepts upon consideration of these areas.

CONCLUSION

Increased sensitivity in tests designed to detect the presence of HBAg has been noted in recent years. The availability of a reliable and safe solid-phase direct radioimmunoassay for the detection of this antigen has resulted in greatly increased protection for patients who must receive blood transfusions in the course of their clinical management. The use of this extremely sensitive test procedure has made possible the diagnosis of long-incubation hepatitis in other patients prior to the onset of clinically recognizable disease. The full impact of the results of this increased sensitivity on the delivery of health care is still to be demonstrated, although few will deny that an extremely important breakthrough has been attained.

REFERENCES

1. Blumberg, B. S.: Polymorphisms of serum proteins and the development of isoprecipitins in transfused patients. *Bull. N.Y. Acad. Med.* **40**:377–386, 1964.
2. Blumberg, B. S., Gerstley, B. J. S., Hungerford, D. A., London, W. T., and Sutnick, A. I.: A serum antigen (Australia antigen) in Down's syndrome, leukemia, and hepatitis. *Ann. Intern. Med.* **66**:924–931, 1967.
3. Maugh, T. H., II: Hepatitis: A new understanding emerges. *Science* **176**:1225–1226, 1972.
4. Krugman, S., Giles, J. P., and Hammond, J.: Infectious hepatitis: Evidence for two distinctive clinical, epidemiological, and immunological types of infection. *J. Amer. Med. Assoc.* **200**:365–373, 1967.
5. Garibaldi, R. A., Rasmussen, C. M., Holmes, A. W., and Gregg, M. B.: Hospital-acquired serum hepatitis. *J. Amer. Med. Assoc.* **219**:1577–1580, 1972.
6. *Summary Report: NHLI's Blood Resource Studies,* DHEW Publication No. (NIH) 73-416. Public Health Service, National Institutes of Health, Bethesda, Md., June 30, 1972, p. 23.
7. Prince, A. M. and Burke, K.: Serum hepatitis antigen (SH), rapid detection by high voltage immunoelectrophoresis. *Science* **169**:593–595, 1970.
8. Alter, H. J., Holland, P. V., Purcell, R. H., et al.: Posttransfusion hepatitis after exclusion of commercial and hepatitis-B antigen-positive donors. *Ann. Intern. Med.* **77**:691–699, 1972.
9. Simmons, H. E., *Amer. Med. News,* July 23, 1973.
10. Ling, C. M. and Overby, L. R.: Prevalence of hepatitis-B virus antigen as revealed by direct radioimmune assay with ^{125}I-antibody. *J. Immunol.* **109**:834–841, 1972.
11. Ginsberg, A. L., Bancroft, W. H., and Conrad, M. E.: Simplified and sensitive detection of subtypes of Australia antigen (HBAg) using a solid-phase radioimmunoassay. *J. Lab. Clin. Med.* **80**:291–296, 1972.
12. Lewis, J. H. and Coram, J. E: Australia antigen detection. Comparison of five CEP and one RIA test systems. *Transfusion* **12**:301–305, 1972.
13. Taswell, H. F.: Incidence of HBAg in blood donors: An overview. In Vyas, G. N., Perkins, H. A., and Schmid, R. (eds.), *Hepatitis and blood transfusion.* Grune and Stratton, New York, 1972, pp. 271–274.
14. Hacker, E. J., Jr., and Aach, R. D.: Detection of hepatitis-associated antigen and anti-HAA. *J. Amer. Med. Assoc.* **223**:414–417, 1973.
15. Sgouris, J. T.: Limitations of the radioimmunoassay for hepatitis B antigen. *N. Engl. J. Med.* **288**:160–161, 1973.
16. Ausria-125, Abbott Laboratories, North Chicago, Illinois.
17. Pike, R. M., Sulking, S. E., and Schulze, M.: Continuing importance of laboratory-acquired infections. *Amer. J. Public Health* **55**:190–199, 1965.
18. Sutnick, A. I., London, W. T., Millman, I., Gerstley, B. J. S., and Blumberg, B. S.: Ergasteric hepatitis: Endemic hepatitis associated with Austrialia antigen in a research laboratory. *Ann. Intern. Med.* **75**:35–40, 1971.
19. U.S. Center for Disease Control. *Hepatitis Surveillance Report No. 33,* 1 Jan. 1971.
20. Marmion, B. P. and Tonlsin, R. W.: Control of hepatitis in dialysis units. *Brit. Med. Bull.* **28**:169–179, 1972.
21. Blumberg, B. S.: Australia antigen. *Amer. J. Med. Technol.* **38**:321–332, 1972.
22. Clark, M. O. and Lewis, J. F.: Incidence of hepatitis-associated antigen among patients in a methadone clinic. *South Med. J.* **66**:389–390, 1973.
23. Davis, L. E., Kalousek, G., and Rubenstein, E.: Hepatitis associated with illicit use of intravenous methamphetamine. *Public Health Rep.* **85**:809–813, 1970.
24. Blumberg, B. S.: Australia antigen—What it means to the hospital clinician. *Resident Staff Physician,* September 1972, pp. 66–84.
25. U.S. Center for Disease Control: *National Nosocomial Infections Study,* Quarterly Report, First Quarter 1972, January 1973.
26. Cossart, Y. E.: Epidemiology of serum hepatitis. *Brit. Med. Bull.* **28**:156–162, 1972.

The Medical Application of the Carcinoembryonic Antigen Assay

CHAPTER SEVEN   JOHN LANGAN, Ph.D.

The relationship between embryonic development and neoplasia was first formulated by Conheim nearly a century ago. The first good experimental evidence was obtained by Erlich and Schöne (1) in 1906. They showed that mice injected with fetal tissue acquired the capacity to reject transplants of tumor tissue which otherwise grew and killed the mice. Adult tissue did not produce this response. In 1930, Hirszfeld and Malber (2) reported the presence of embryonal antigens in extracts from tumors. The subject was not taken seriously until the 1950s and early 1960s. The original observation of what is now called carcinoembryonic antigen (CEA) was made by Gold and Freedman in 1965 (3). They prepared extracts of human colonic adenocarcinoma, raised antisera against them in rabbits, and then tested the antisera by immunoprecipitation. All tumor extracts contained antigens not present in normal colonic tissue from the same individuals. It was later found that the same antigens were present in smaller amounts in primary carcinomas in other areas of the digestive tract. The antigens were absent from benign tumors and polyps, and from nonneoplastic diseases of the digestive tract such as ulcerative colitis and diverticulitis. However, they were found in the the intestine, liver, and pancreas of human fetuses during the first two trimesters, though not in the corresponding tissues in the normal adult.

CEA was subsequently purified and found to be a closely related group of water-soluble, 200,000 molecular weight glycoproteins with no glucose or N-acetyl-D-galactosamine in their side chains, and no methionine or cystine in their peptide chains. The glycoproteins are localized in cell membranes, and particularly in the glycocalyx.

Gold and his colleagues developed a radioimmunoassay for CEA, and performed the first clinical trial. Unfortunately, they found CEA in the serum of 35 of 36 patients with adenocarcinoma of the large intestine, but not in sera from normal individuals or from patients with either nonmalignant diseases of the digestive tract or malignant diseases in other areas of the body (4). Further studies have not confirmed the localization of the release of CEA into the blood to be restricted to carcinomas of the large intestine. Martin and Martin (5) found low concentrations of CEA in normal adult colon. Moore et al. (6) found CEA in serum of patients with carcinoma of the upper digestive tract, pancreas, and bronchi, as well as with carcinoma of the colon.

Using a more sensitive radioimmunoassay than Gold, in which the antiserum may be reacting with a different, ion-sensitive antigenic site as the CEA molecule, Lo Gerfo et al. (7) found CEA in the serum of patients with primary carcinomas in widespread areas of the body. Reynoso et al. (8) have extended this work, and we would like to present our own experiences with the CEA assay. We have performed 3000 assays on 2400 patients, over 400 of whom were studied serially on from two to six occasions during a period of 18 months. The assay used in our study was developed by Hansen (7) and measures down to 0.5 ng/ml with a reproducibility of ±0.5 ng/ml. The assay requires extraction of 0.5 ml of patient plasma with perchloric acid, removal of unconjugated proteins by centrifugation, and removal, by overnight dialysis, of the perchloric acid. The dialysate is then incubated with goat anti-CEA, followed by a second incubation after the addition of a known amount of purified CEA labeled with iodine-125. The antigen-antibody complex is separated from the remaining unbound CEA by the addition of zirconium phosphate gel. A standard curve in which purified CEA replaces the plasma extract is prepared to quantitate the CEA in the patient specimen.

Table 1 shows a general summary of the results. Columns 2, 3, and 4 list the number of specimens with less than 2.5, 2.5 to 5, and over 5 ng/ml of CEA. Column 5 lists the percentage of patients with abnormally high (or positive) levels of circulating CEA, that is, those with more than 2.5 ng/ml. The patients are

Table 1 Summary of the Plasma CEA Levels and Percentage of Abnormally High CEA Levels in Five Classes of Patients

Cancer	<2.5 ng/ml	Gray	>5.0 ng/ml	Percent +
1. Lung	5	11	39	91
Upper gastrointestinal	6	10	19	83
Lower gastrointestinal	9	7	32	81
Brain and central nervous system	0	3	12	100
Obstetrical-gynecological	8	4	9	62
Prostate	2	1	5	75
Bone metastases	2	3	16	90
Total	32	39	132	84
2. Head and neck	70	37	16	43
Breast	23	6	13	45
Skin	29	3	5	28
Total	122	46	34	40
3. Hodgkin's disease, leukemias, malignant melanoma, myeloma, and sarcomas	29	10	10	41
4. Nonmalignant	264	31	23	17
5. Normal	200	0	0	0

Table 2 Plasma CEA Levels in Patients with Benign Lesions

Benign lesion	<2.5 ng/ml	Gray	>5.0 ng/ml
Breast	48	1	0
Obstetrical-gynecological	17	0	1
Colon plus rectum	6	0	1
Head and neck	7	0	0
Prostate	10	1	0
Lipomas	6	0	0
Miscellaneous	27	4	0
Total	121	6	2

divided into five groups, the first group having cancers that release abnormally high levels of CEA. The second group of patients had cancers at sites that usually do not release abnormally high levels of CEA. The third group had malignant diseases whose initial diagnosis could best be made by present conventional methods without the help of this test. Very preliminary data suggest that this third group may have released abnormally high levels of CEA only during active phases of the disease and not during periods of remission. Information on the degree of activity of these diseases may be of much greater value to the therapist than assistance in the initial diagnosis.

The fourth group of patients was admitted to Temple University Hospital with severe medical or surgical problems, but with no suspicion of malignant disease, and yet 17% had positive levels (above 2.5 ng/ml of CEA). Tables 2 and 3 indicate that this may not be as serious a problem in using the test as the simplified summarized figures initially suggest. At first sight this number of 17% is exorbitantly high but, as will be seen, these patients all suffered from diseases most of which had inflammation as a prime symptom. In contrast, not one of the 200 apparently normal healthy volunteers in group 5 showed levels that were sustained at over 2.5 ng/ml. Two of these volunteers had initial levels in the "gray" zone, which on repeat sampling 1 month later were shown to have returned to below 2.5 ng/ml. Thus at the onset it must be stressed that normal patients have at least a very high

Table 3 Plasma CEA Levels in Patients with Inflammatory Diseases

Inflammatory	<2.5 ng/ml	Gray	>5.0 ng/ml
Upper gastrointestinal	11	3	1
Lower gastrointestinal	33	4	1
General	31	5	3
Infectious	5	2	5
Liver cirrhosis	2	7	4
Total	82	21	14

probability of having normal levels of CEA and that the apparent false positives of category 4 are patients who were mostly in acute stages of serious illness. The likelihood of wrongly diagnosing a healthy patient by this test is not great. To decrease this probability of error even further, we have chosen what we call a gray zone between 2.5 and 5.0 ng/ml. Patients whose plasma CEA levels fall in the gray zone receive repeat assays at fixed time intervals to observe an increase or decrease in the CEA level. Eighty-five percent of all patients with levels of over 2.5 ng/ml, and 90% of those over 5.0 ng/ml, have already been diagnosed as suffering from malignant disease.

When the high CEA level is caused by newly regenerating cells in the healing process of inflammatory conditions, the plasma CEA levels may be temporarily high and then return to normal, that is, below the 2.5 ng/ml, when the inflammation heals. Most such temporary high levels are in the gray zone, and only very infrequently are over 10 ng/ml. To use one or two marginally high CEA levels as definitive confirmation of the presence of a carcinoma when the patient is known to have other conditions that could cause inflammation is not advisable. However, we have seen levels as high as 2000 ng/ml in cases of carcinoma of the colon, lung, or liver. It may already be safe to suggest that levels of about 20 ng/ml can be due only to carcinoma. In fact, nearly all patients with levels of over 20 ng/ml have already been shown to have cancers with metastases.

Table 2 shows a summary of the results for patients whose primary medical problem was inflammatory disease, to which we attribute the abnormally high level of plasma CEA. The highest level of CEA in any of these patients was 8 ng/ml of CEA. Later follow-up in these patients revealed levels that had returned to normal. For this discussion we consider alcoholic cirrhosis as an inflammatory disease. Heavy drinkers and heavy smokers may well be the greatest source of diagnostic problems. Inflammation of the liver and lungs due to excessive drinking and smoking, respectively, may produce slightly elevated levels of CEA from normal regenerating cells. Unfortunately, the real problem is that organs so abused are rendered more susceptible to the growth of malignant tissue. Thus the physician will have difficulty in deciding whether the CEA is from regenerating normal tissue or from malignant cells. Schwarz and Fleischer (10), at Memorial Hospital for Cancer and Allied Diseases, New York, are making an extensive study of the correlation of plasma CEA levels and smoking. Some workers in this field state that a sustained high level in heavy smokers is caused only by undiscovered malignant cells. Gold (9), at Montreal General Hospital, had at one time seen 81 patients with abnormally high plasma CEA levels who had, at the time of assay, no diagnosed malignant disease. Subsequently, 39 of these patients were diagnosed as having a malignancy. It may well be assumed that many if not all the others may yet be found to have an undiscovered malignancy.

Table 3 summarizes the results in patients with benign lesions. Of 127 patients only 9 had abnormally high levels, only 3 of which were over 5 ng/ml. The absence of malignant disease has not been definitively ruled out in any of these patients with positive levels, who are tentatively diagnosed as having only benign lesions or inflammatory conditions.

Table 4 gives a comparison between CEA levels in patients with benign and malignant lesions, and also between patients with inflammatory conditions and malignant lesions. Once again it must be stressed that because we have not found

Table 4 Plasma CEA Levels in Patients with Benign Lesions in Comparison to the Levels Found in Patients with Malignant Lesions at the Same Anatomical Sites[a]

Cancer site	Benign			Malignant		
	<2.5 ng/ml	Gray	>5.0 ng/ml	<2.5 ng/ml	Gray	>5.0 ng/ml
Breast	48	1	0	No mx 20	3	1
				mx 3	3	12
Prostate and bladder	12	1	0	2	1	5
Obstetrical-gynecological	17	0	1	8	4	9
Lower gastrointestinal	6	0	1	9	7	32

Cancer Site	Inflammatory			Malignant		
	2.5 ng/ml	Gray	5.0 ng/ml	2.5 ng/ml	Gray	5.0 ng/ml
Upper gastrointestinal	11	3	1	6	10	19
Lower gastrointestinal	33	4	1	9	7	32

[a] Also a comparison between plasma CEA levels in patients with inflammatory diseases and with malignant diseases.

metastases does not mean that they do not exist, just as the fact that we have not found a malignancy in any of the cases in group 4 in Table 1 does not mean that a malignancy does not exist. In most of those cases no specific attempt was even made to find a malignancy for such reasons as the very advanced age of the patient, the lack of overt symptoms of cancer, and the conclusive diagnosis of the disease for which the patient was admitted. Some of our physicians are now becoming sufficiently experienced with the CEA assay that investigations for cancer are being undertaken when a sustained high CEA level is discovered. With this in mind the data being presented may be doing these tests a severe injustice by making the number of apparent false positives unnecessarily high. This could be due to a lack of a definitive data on unsuspected malignancies which may be present.

The data on carcinoma of the breast is of particular interest. Forty-six patients with benign lesions had normal levels of CEA. The patient with a level in the gray zone had massive lesions of the breast in which malignant cells have as yet not been found. Similar data are seen in the patients with malignancies restricted to the breast. Nineteen patients had normal levels, three had levels in the gray zone, and one had a level above 5 ng/ml. However, when carcinoma of the breast metastasizes to another organ, usually lung or liver, the plasma CEA usually increases to abnormally high levels (several hundred nanograms per milliliter). Although plasma CEA levels are apparently of little help in the initial diagnosis of breast cancer, they will probably be of great assistance to the surgeon in deciding whether the cancer is restricted to the breast or whether it has already metastasized. The type and extent of surgery may thus be affected. Fortunately, the three sites at which the CEA test is often not indicated, namely, breast, skin and neck, and throat and mouth, are all on

Table 5 Plasma CEA Levels in Patients with a Variety of Malignant Diseases[a]

Disease	<2.5 ng/ml	Gray	>5.0 ng/ml
Hodgkin's disease	7	3	4
Leukemias	11	3	3
Malignant melanomas	4	1	2
Multiple myelomas	2	1	1
Sarcomas	5	2	0

[a] In these diseases the test may be of limited use for initial diagnosis, but may be useful in monitoring the activity of the disease, and thus assist in determination of the degree of intensity of therapy.

or near the surface. Thus the patient himself may detect an abnormality such as a lump or abrasive area irritation much earlier than he would in the case of many deep-seated cancers such as bone, lung, or liver.

High levels of CEA are found in only 30–40% of patients suffering from such diseases as sarcoma, multiple myeloma, malignant melanoma, leukemias, and Hodgkins disease. The initial data presented in Table 5 do not adequately show the whole picture. Many of these cases have been studied serially for many months, during which time the CEA levels in some patients have changed. A rise in the CEA level has in many cases been accompanied by the clinical diagnosis of an increase in the degree of activity of the diseae. Subsequent diagnosis after the intensification of therapy has also paralleled the patient's return to a stage of remission.

Many cases of mycosis fungoides have been followed in which the CEA levels remained normal while the disease was restricted primarily to the skin. In three cases in which the disease rapidly flared into a widespread systemic lymphoma, the level of CEA rose equally rapidly to abnormal high levels. These three cases were terminal, and the change in the CEA level did not come early enough to warn the physician of the impending fatal invasions of the lympathic system. However, even with prior warning it is doubtful that any present therapy would have saved the patient. As stated already, if these initial results are confirmed, the test may be very useful in monitoring the degree of activity of disease, and thus be of more value to the physician than merely another test to assist him in early diagnosis.

One area in which the test has already been shown to be of immediate and clear-cut value is in the monitoring of the completeness of surgery. Some individual cases of note are shown in Table 6. Patients with abnormally high levels of plasma CEA before surgery should be followed with periodic CEA levels after surgery. If the CEA level remains high, it could mean that not all the malignant tissue was removed from the site of the operation, or that metastases existed elsewhere. A return of the level to normal, and a later increase, suggest either a recurrence of the original cancer, a metastasis, or new primary cancer elsewhere.

Table 7 shows our anomolous results. K.D. is a 46 year old woman of normal weight who smoked three packs of cigarettes a day for 20 years. She has had two apparently benign lesions, of the breast removed within the last 2 years. Her CEA

Table 6 Plasma CEA Levels before Removal of a Carcinoma and at Intervals after Surgery

	CEA (ng/ml)		
		Postoperative	
Cancer site	Pre-operative	4 days	Later
Cervix	8	3	2
Colon	8	6	1
Colon	9	2	
Cecum	12	8	5
Lung	135	—	5, 3
Lung	4	—	2, 1
Mouth	7	3	
Larynx		4	3, 3, 1, 2, 2
Neck plus Larynx	2	2	2

level was usually around 13 ng/ml, but had been as high as 20 ng/ml. She stopped smoking at the same time that she had some of the plantar warts removed from her feet. Her CEA level dropped to 5 ng/ml within 6 weeks. Her remaining plantar warts are being treated medically, and her CEA level is now down to 4 ng/ml after 3 months.

W.D. is a 54 year old male sales executive who is 40 lb overweight, has smoked three packs of cigarettes a day for at least 20 years, and had "indigestion." His CEA level is 5 ng/ml. E.D. had a CEA level of 4 and later of 5. She does not qualify for the category of "normal healthy adult." R.A. is alcoholic and suffers from nephrosis. He was maintained in a hospital for 3 months, during which time the inflammatory tissue in his kidney was greatly reduced and his alcoholic intake was at least restricted. The patient's CEA level dropped from 7 to 1 ng/ml.

In summary, this procedure is ready to be performed in competent laboratories

Table 7 Patients with Abnormally High CEA Levels Who Do Not Fit into Any of the Previous Categories

Patient	Age	Weight	Cigarettes smoked (packs/day)	Abnormalities	CEA (ng/ml)
K.D.	46	Normal	3	Plantar warts	13–20
	47	Normal	0	—	5
W.D.	53	+40 lb	3	"Indigestion"	5
E.D.	63	+50 lb	1	Diabeties	4, 5
R.A.	53	?		Alcoholism, nephrosis	7
	53	?		Dry 3 months	1

with well-trained personnel who are willing to perform an assay of this degree of delicacy and requiring this amount of care with extreme quality control. The *results* of the tests are ready for physicians who are willing to completely understand the strengths and weaknesses that have so far been demonstrated. The test can be used in confirmation of a suspected carcinoma, for differentiation between malignant and benign growths, and between malignant diseases and inflammatory conditions, and for the monitoring of therapy be it radiological, chemotherapeutic, or surgical.

REFERENCES

1. Schone, G.: Untersuchungen uber karzinomimunitat bei mausen. *Muench. Med. Wochenschr.* **53**:2517–2519, 1906.
2. Hirszfeld, L. and Malher, W.: Ueber krebsanoekoerper bei krebs kranken. *Klin. Wochenschr.* **9**:342, 1930.
3. Gold, P. and Freedman, S. O.: Demonstration of tumor-specific antigens in human colonic carcinomata by immunological tolerance and absorption techniques. *J. Exp. Med.* **121**:439, 1965.
4. Thompson, D. M. P., Krupey, J., Freedman, S. O., and Gold, P.: The radioimmunoassay of circulating carcinoembryonic antigen of the human digestive system. *Proc. Nat. Acad. Sci. U.S.* **64**:161, 1969.
5. Martin, F. and Martin, M. S., Demonstration of antigens related to cancer. *Int. J. Cancer.* **6**:352, 1970.
6. Moore, T., Dhar, P., Marcon, N., Moore, D. L., Kupchik, H. Z., and Zamcheck, N., Carcinoembryonic antigen assay in cancer of the colon and pancreas, and other digestive diseases. *Amer. J. Dig. Dis.*, **16**:1, 1971.
7. LoGerfo, P., Krupey, J., and Hansen, H. J.: Demonstration of a common antigen with neoplasia assay using zirconyl phosphate gel. *N. Engl. J. Med.* **285**:138, 1971.
8. Reynoso, G., Chu, T. M., Holyoke, D.: Carcinoembryonic antigen in patients with different cancers. *J. Amer. Med. Assoc.* **220**:361, 1972.
9. Gold, P. Discussion at the Second Annual Symposium of Carcinoembryonic Antigen, Department of Medical Research, Hoffman-LaRoche, Inc., Nutley, New Jersey.

Radioimmunoassay of Gastrin

CHAPTER EIGHT JAMES E. McGUIGAN, M.D.

The production and characterization of antibodies to gastrin have now made possible the sensitive and specific radioimmunoassay measurement of this gastrointestinal hormone (1–7). Gastrin is a polypeptide, or more precisely, a group of structurally related polypeptides, found principally in the mucosa of the gastric antrum and upper small intestine. In 1964, Gregory and Tracy (8–9) identified the structures of two gastrins they isolated from porcine antral mucosa. These were heptadecapeptides (each containing 17 amino acid residues in a single polypeptide chain) and differed only by sulfation of the tyrosyl residue in position 12. The sulfated form of heptadecapeptide gastrin has been designated gastrin II, and the nonsulfated form gastrin I (Figure 1). The availability of these pure heptadecapeptides in synthetic and naturally occurring forms has contributed immensely to the development of radioimmunoassay techniques for measurement of gastrin.

The primary action of gastrin has been viewed as its capacity to stimulate gastric acid secretion (10), and it is in relation to this capacity to stimulate acid secretion that radioimmunoassay of gastrin finds its greatest applicability. However, as with most other polypeptide hormones, gastrin exhibits a multiplicity of activities, action having been demonstrated on various components of the gastrointestinal tract as well as on other endocrine organs. These include, among others, the capacity to stimulate pancreatic bicarbonate secretion, to raise motor activity of the large and small intestine, to increase the tone of the lower esophageal sphincter, to decrease the tone of the pyloric sphincter, to promote hepatic bile flow, to stimulate insulin release, to stimulate gastric secretion of pepsin and intrinsic factor, and to stimulate calcitonin release from the thyroid gland. It remains to be established which of all the multiple activities of gastrin are pharmacological, and which represent physiological activities. It appears reasonably well established that gastrin is of great physiological importance in control of the rate of gastric acid secretion and probably, under physiological circumstances, represents an important component in the maintenance of lower esophageal sphincter pressure.

Gastrin release into the circulation is stimulated by a variety of factors including antral distention, vagal and other cholinergic stimulation, feeding—particularly of a

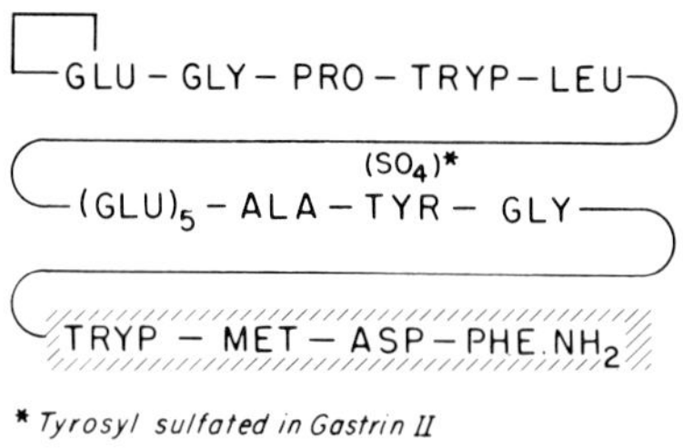

```
    ┌─GLU-GLY-PRO-TRYP-LEU─┐
       (SO₄)*
    ─(GLU)₅-ALA-TYR-GLY─
    ─TRYP - MET-ASP-PHE.NH₂
* Tyrosyl sulfated in Gastrin II
```

Fig. 1. Amino acid sequence of human gastrin heptadecapeptide.

protein-containing meal—and ingestion of amino acids, particularly glycine (11,12). In addition, gastrin release may be promoted by hypercalcemia secondary to intravenous administration of calcium, and by oral ingestion of calcium in amounts insufficient to alter circulating calcium concentrations (13–15).

METHODS OF PRODUCING ANTIBODIES TO GASTRIN

An assortment of techniques has been developed and applied for the successful production of antibodies to gastrin for use in radioimmunoassay (1–7). For the most part, two principal methods are currently being utilized to stimulate production of antibodies to gastrin. The first method entails the use of human gastrin I heptadecapeptide as antigen, and is described as follows. Human gastrin I residues 2 to 17 (SHG:2–17) are covalently conjugated to bovine serum albumin by use of 1-ethyl-3-(3-dimethylaminopropyl) carbodiimide. By utilizing this technique, on the average, 8.4 peptide groups have been coupled to each bovine serum albumin molecule of molecular weight 68,000. This method effectively enhances the limited immunogenicity of the gastrin heptadecapeptides by utilizing human gastrin I as the haptenic determinant. The peptide–protein conjugate is administered to randomly bred New Zealand white rabbits in complete Freund's adjuvant by footpad injection at two or three intervals of 1 month, with the production of antibodies suitable for gastrin radioimmunoassay produced in the rabbits following the third and subsequent immunizations. The second principal technique for producing antibodies for radioimmunoassay of gastrin is that of utilizing as antigen partially purified porcine gastrin (16). In utilizing this method from 3 to 60 mg of partially purified gastrin emulsified in Freund's adjuvant is injected subcutaneously on the inner aspect of the thigh of guinea pigs. Antibodies derived by both these techniques have been found suitable for sensitive and specific radioimmunoassay measurement of gastrin.

An additional technique for producing antibodies for radioimmunoassay of gastrin has involved immunization of rabbits with the carboxyl-terminal tetrapeptide amide of gastrin or pentagastrin (which contains the carboxyl-terminal tetrapeptide amide of gastrin) covalently conjugated to bovine serum albumin by use of carbodiimide (17,18). The carboxyl-terminal tetrapeptide amide of gastrin, which may be viewed as the active-site region of gastrin, exhibits all the physiological properties of the intact gastrin molecule, although on a molar basis it is only from one-sixth to one-tenth as potent as the intact heptadecadeptide. Antibodies to gastrin produced in this manner exhibit antibody binding of gastrin, but are of limited usefulness because of their lack of specificity, particularly in respect to their cross-reactivity with cholecystokinin-pancreozymin (18,19). Cholecystokinin-pancreozymin is a gastrointestinal hormone residing in the mucosa of the upper small intestine, which contains 33 amino acid residues and possesses the same carboxyl-terminal pentapeptide amide sequence as does gastrin (20). For this reason antibodies to the tetrapeptide amide portion of the gastrin molecule exhibit approximately equivalent immunological reactivity with both gastrin and cholecystokinin-pancreozymin (19). Therefore, antibodies to the tetrapeptide amide of gastrin are not suitable for the specific measurement of the polypeptide hormone gastrin. Antibodies to intact, or virtually intact, gastrin molecules exhibit minimal binding of cholecystokinin-pancreozymin when compared with their binding of gastrin molecules (16,21). Antibodies to human gastrin I exhibit more apparent antibody specificity for, and reactivity with, the smaller tetrapeptide amide portions of the gastrin molecule than for the larger remaining portion of gastrin which contains the first 13 amino acid residues of the peptide (amino tridecapeptide) (21).

SEPARATION OF ANTIBODY-BOUND FROM ANTIBODY-FREE GASTRIN

Each currently available radioimmunoassay technique depends upon the application of an appropriate method for separating antibody-bound from antibody-free radiolabeled hormone, in this instance the polypeptide hormone gastrin. Among others, the following methods have been successfully applied to separate antibody-bound from antibody-free gastrin. On the basis of differences in molecular size, Sephadex gel filtration has been used to separate antibody-bound from unbound radiolabeled gastrin (or gastrin tetrapeptide) (16,17). The "double-antibody technique" has been used to precipitate radiolabled gastrin bound by rabbit antibodies to gastrin by the addition in excess of precipitating goat antibodies to rabbit gamma globulin (1,22,23). Antibody-bound radiolabeled gastrin has been precipitated by using ethanol, leaving nonprotein-bound gastrin in the supernatant solution (2,24). Gastrin, and many other small organic molecules, can be readily adsorbed to the surface of dextran- or albumin-coated charcoal particles, antibody gastrin remaining in the supernatant solution (4,25). Insoluble anion-binding resins, for example, the Amberlite resin CG-4B, can be used to complex with gastrin molecules, which, by virtue of their generous content of dicarboxylic glutamic acid residues, possess a strong net negative charge, thereby separating them from antibody-bound gastrin remaining in the supernatant solution (16).

We have found that the use of the anion resin CG-4B possesses distinct advantages over the use of double-antibody systems. Of the various methods for separation of antibody-bound from antibody-free radiolabeled gastrin, we prefer the use of CG-4B resin. The anion-binding resin CG-4B may be used with antibodies to human gastrin I in the following manner. Incubations are performed in duplicate or triplicate in 10 × 75 mm disposable glass tubes with incubation volumes adjusted to a final volume of 2.5 ml by addition of 0.02 barbital buffer (pH 8.4) containing ovalbumin 2.5 mg/ml (EA-BARB). Human [^{125}I]gastrin I (residues 1 to 17), approximately 6000 cpm, in 0.2 ml EA-BARB is included. The incubation mixture contains 0.2 ml of antiserum to human gastrin I, which is diluted sufficiently in EA-BARB so that a bound/free ratio of human [^{125}I]gastrin I is achieved that is between 0.6 and 1.0. The final dilution of the antiserum to gastrin in the incubation mixture is usually from 1:500,000 to 1:1,500,000. Known quantities of human gastrin I for the construction of calibration curves, as well as unknown samples, for example, serum or tissue extracts, in appropriate dilutions, in 2 ml of incubation mixture are included. Immune incubations are conducted at 4°C for periods from 2 to 5 days. (When antibodies of sufficiently high affinity are used, this incubation period may be even further shortened.) Following completion of incubation the CG-4B resin may be readily used to separate antibody-free from antibody-bound human [^{125}I]gastrin I. CG-4B at a concentration of 100 mg/ml is suspended in 0.02 M barbital buffer (pH 8.4), and 0.2 ml of the CG-4B-containing suspension is added to each incubation tube, after which the tube is gently mixed for 5–10 seconds using a vortex mixer. The tubes are then centrifuged for 3 minutes at 1000 × G at 4°C; antibody-bound human [^{125}I]gastrin I remains in the supernatant, and antibody-free [^{125}I]gastrin I which is bound to the CG-4B particles is readily precipitated. Calibration diagrams (Figure 2) may be established using bound/free ratios with subsequent application of these calibration diagrams for measurement of gastrin contained in samples of unknown gastrin concentration. Several characteristics of the behavior and use of CG-4B must be emphasized. A pH of 8.4 represents an ideal pH for the activity of CG-4B in binding gastrin. This resin is not nearly as effective at lower pH, for example, pH 7 or 7.4. The maximum ratio of serum to total incubation volume should not exceed 10%. Binding of gastrin by CG-4B, and thereby its effectiveness in separation of antibody-bound from antibody-free radiolabeled gastrin, is diminished by increases in molar concentration that sub-

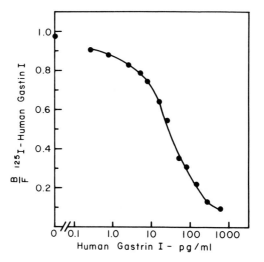

Fig. 2. Calibration curve demonstrating ratios of antibody-bound human [^{125}I]gastrin I to antibody-free human [^{125}I]gastrin I in the presence of varying amounts of nonradiolabeled human gastrin I. Antibodies to human gastrin I (SGH:2–17) were utilized at a final dilution of 1:500,000. Values expressed on the horizontal axis refer to picograms of gastrin per milliliter of incubation mixture; sensitivity level, approximately 0.3 pg/ml.

stantially exceed those obtained with the system as described above. Finally, inclusion of large amounts of heparin (like gastrin, a strongly negatively charged molecule) inhibits CG-48 binding of gastrin, and may thereby result in lower than true estimations of gastrin content. This observation has led us to perfer the measurement of gastrin in serum rather than in heparinized plasma when utilizing the CG-4B separation technique.

VARIOUS SPECIES OF GASTRIN

The polypeptide hormone gastrin is present both in serum and in tissue extracts in various immunoreactive gastrin species (1,2,26–29). The form in which the structure of gastrin was originally characterized was that of the heptadecapeptide gastrins, as purified and described by Gregory and Tracy (1,2). In the heptadecapeptide species of gastrin, the amino terminus is blocked by a pyroglutamyl residue, and the carboxyl terminus by amidation (Figure 1). Gastrin II has been distinguished from gastrin I by the presence of sulfation of the tyrosyl residue in position 12, whereas this is nonsulfated in gastrin I (Figure 1). Gastrins I and II in mammalian species appear to be approximately equal in their biological activity. It has been shown that the major circulating immunoreactive species of gastrin is a molecule that is larger than the heptadecapeptide gastrin (containing 34 amino acid residues), and which possess a less pronounced net negative change than heptadecapeptide gastrin; this form(s) has been designated "big" (or "basic") gastrin (26,27). Big gastrin, when incubated with trypsin, yields a fragment with the amino acid composition and biological activity of heptadecapeptide gastrin. It is probable therefore that big gastrin represents a larger polypeptide molecule, perhaps a variety of precursor form of heptadecapeptide gastrin. Big gastrin, as well as heptadecapeptide gastrin, appear to exist in both sulfated and non sulfated forms. Although the major circulating form of gastrin is big gastrin, the predominant species of gastrin in the antral gastric mucosa is heptadecapeptide gastrin (27). In addition to the antral mucosa, gastrin is also found in the mucosa of the upper small intestine. The amount of gastrin present in the duodenal and jejunal mucosa gradually decreases with caudad progression. However, the ratio of immunoreactive gastrin, in the form of big gastrin, to that of heptadecapeptide gastrin increases with progression down the gastronintestinal tract. There is evidence that big gastrin has a longer biological half-life than heptadecapeptide gastrin (30). Recently, Yalow and Berson identified a substantially larger species of gastrin, designated "big, big" gastrin, which appears in the void volume with Sephadex G-50 gel filtration (29). Incubation of big, big gastrin with trypsin also appears to yield heptadecapeptide gastrin. Presently, available information supports the view that heptadecapeptide gastrin is the biologically active structural component of these larger gastrin polypeptide species (both big gastrin and big, big gastrin).

Antibodies utilized in radioimmunoassay, as previously described following immunization with either human gastrin I residues 2 to 17 to partially purified porcine gastrin, exhibit immunoreactivity with heptadecapeptide gastrins, big gastrin, and big, big gastrin molecules. Most antibodies to gastrin utilized in radioimmunoassay exhibit approximately equilivalent immunological reactivity with gastrin I and gastrin II species (1,16,21,31).

APPLICATIONS OF ANTIBODIES IN THE RADIOIMMUNOASSAY MEASUREMENT OF GASTRIN

Radioimmunoassay of gastrin has been used in a variety of studies measuring serum and plasma concentrations of this gastrointestinal hormone, as well as gastrin concentration in tissue extracts (2,16,32–39). At the present time the major practical clinical application of gastrin radioimmunoassay is the detection of high concentrations of gastrin in serum or plasma of patients with the Zollinger-Ellison syndrome (32). As is well recognized, the Zollinger-Ellison syndrome is characterized by the presence of nonbeta islet cell tumors of the pancreas, extremely severe ulcer disease of the upper gastrointestinal tract, and usually greatly elevated rates of gastric acid secretion. It has now been well established that these nonbeta islet cell tumors of the pancreas are rich in gastrin, that hypergastrinemia is characteristic of the Zollinger-Ellison syndrome, and that the pathophysiological sequelae of the Zollinger-Ellison syndrome result from the high circulating concentrations of gastrin in the blood. Normal fasting serum gastrin concentrations in most studies fall within the range from immeasurably low (<10 pg/ml) to 200 pg/ml, with means averaging approximately 75 pg/ml. In contrast, fasting serum concentrations in patients with the Zollinger-Ellison syndrome are usually greater than 300 pg/ml and may be as high as 350,000 pg/ml. Patients with common duodenal ulcer disease, even in the face of marked gastric acid hypersecretion, in contrast to Zollinger-Ellison patients, do not have increased fasting serum gastrin levels (38). Interesting relationships have been identified between calcium and serum gastrin concentrations, so that the calcium infusion test represents an important test for the verification of the diagnosis of the Zollinger-Ellison syndrome. Intravenous calcium infusion provides a potent stimulus to gastrin release in patients with Zollinger-Ellison tumors (13–15). The increase in gastrin concentration with calcium infusion in patients with Zollinger-Ellison tumors usually substantially exceeds that observed in patients with usual peptic ulcer disease, as well as in normal control subjects (14,15). Calcium is administered by constant intravenous infusion as calcium gluconate (5 mg calcium/kg hour for a 3-hour period). Serum gastrin concentrations are obtained prior to intravenous calcium infusion and at 20 minute intervals for 4 hours following initiation of intravenous calcium infusion. In patients with Zollinger-Ellison tumors, serum gastrin concentrations increase substantially, with increments almost always exceeding 500 pg/ml. Conversely, although increases in serum gastrin concentrations are observed in normal individuals and in patients with common duodenal ulcer disease, these increases are usually much smaller than those observed in Zollinger-Ellison patients, often approximate 200 pg/ml, and seldom exceed 500 pg/ml.

A second provocative test for the Zollinger-Ellison syndrome is that of intravenous secretin infusion. In normal individuals, as well as in patients with common duodenal ulcer disease, intravenous secretin administration decreases rate of gastric acid secretion, which appears to be achieved by inhibition of gastrin-mediated gastric acid secretion and by inhibition of gastrin release into the circulation (40,41). In many patients with the Zollinger-Ellison syndrome, a paradoxical increase in serum gastrin concentration is observed following intravenous administration of secretin (42,43). Secretin has been administered in doses of 1 unit per kilogram body weight given by a single intravenous injection (43), and by progressive doses up to 9 units/kg hour administered by constant intravenous infusion (42). The promotion of substantial increases in serum gastrin concentration by calcium infusion appears to be a virtually constant characteristic of patients with the Zollinger-Ellison syndrome. Increases in serum gastrin concentration following secretin administration initially were thought to represent unusual manifestations in selected patients with the Zollinger-Ellison syndrome, however, with increasing experience, it appears that many, and perhaps all, patients with Zollinger-Ellison tumors respond by secretin administration with enhanced gastrin release. The absolute frequency and reliability of detection of substantial increases in serum gastrin concentration with secretin administration in confirmation of the diagnosis of the Zollinger-Ellison syndrome remain to be established. Thus a serum gastrin profile may be established in Zollinger-Ellison patients, which has the following characteristics: (1) increased fasting serum gastrin concentrations, (2) substantial increases in serum gastrin concentration following calcium infusion, and (3) the frequently observed phenonmenon of increases in serum gastrin concentration following secretin infusion.

RADIOIMMUNOASSAY OF SERUM GASTRIN IN PATIENTS WITH PERNICIOUS ANEMIA, CHRONIC GASTRITIS, AND GASTRIC ATROPHY

In addition to the recognition that hypergastrinemia characterizes the Zollinger-Ellison syndrome, it has now been established that elevated fasting serum gastrin concentrations are also found in most (but not all) patients with pernicious anemia (16,34). As is well known, patients with pernicious anemia have gastric mucosal atrophy with achlorhyria. The marked increases in serum gastrin concentration found in patients with pernicious anemia appear to result from two causes. First, under normal circumstances, gastrin release from the antral mucosa is effectively inhibited by intragastric hydrogen ion; when the pH within the stomach is reduced to 3 or less, gastrin release begins to be inhibited, and when the pH is reduced to 1.5 or less, gastrin release is eliminated (44–46). In patients with pernicious anemia, even with maximum histamine or histalog stimulation, the pH remains above 6. Thus, in patients with pernicious anemia, the normal mechanism for inhibition of gastrin release with increasing hydrogen ion concentration within the stomach is absent, thereby permitting excessive gastrin release into the circulation. This hypothesis is supported by the observation that placement of 0.1 N hydrochloric acid within the stomach of patients with pernicious anemia reduces plasma gastric concentrations toward normal (16). An additional mechanism that may contribute to these increases in serum gastrin concentration may involve the observation by some groups of investigators of an increased frequency of gastrin-containing cells within the gastric mucosa of patients with pernicious anemia (47,48).

SUMMARY

By utilizing a variety of techniques it has been possible to produce antibodies suitable for radioimmunoassay measurement of the polypeptide hormone gastrin. A variety of techniques has been applied for satisfactory radioiodination of gastrin, as well as separation techniques to implement radioimmunoassay measurement of gastrin. The most important clinical application of radioimmunoassay measurement of gastrin at the present time is identification of increased gastrin levels in patients with the Zollinger-Ellison syndrome. These patients may be further identified and characterized by calcium infusion and secretin infusion studies. Gastrin exists in a variety of molecular species, each of which exhibits imunoreactivity with antibodies currently in use in radioimmunoassay of gastrin. It is anticipated that radioimmunoassay measurement of gastrin will contribute substantially to our understanding of the behavior of this polypeptide hormone in man in health and disease.

REFERENCES

1. McGuigan, J. E.: Immunological studies with synthetic human gastrin. *Gastroenterology* **54**:1005–1011, 1967.
2. Stremple, J. F. and Meade, R. C.: Production of antibodies to synthetic human gastrin I and radioimmunoassay of gastrin in the serum of patients with the Zollinger-Ellison syndrome. *Surgery* **64**:165–174, 1968.
3. Charters, A. C., Odell, W. D., Davidson, W. D., and Thompson, J. C.: Gastrin: Immunochemical properties and measurement by radioimmunoassay. *Survey* **66**:104–110, 1969.
4. Hansky, J., and Chain, M.D.: Radioimmunoassay of gastrin in human serum. *Lancet* **2**:1388–1390, 1969.
5. Ganguli, P. C. and Hunter, W. M.: A radioimmunoassay for gastrin. (abstract) *Gut*, **10**:413, 1969.
6. Trudeau, W. L., McGuigan, J. E.: Serum gastrin levels in patients with peptic ulcer disease. *Gastroenterology*, **59**:6–12, 1970.
7. Rehfeld, J. F., Stadil, F., and Rubin, B.: Production and evaluation of antibodies for the radioimmunoassay of gastrin. *Scand. J. Clin. Lab. Invest.*, **30**:221–232, 1972.
8. Gregory, R. A., and Tracy, H. J.: The constitution and properties of two gastrins extacted from hog antral mucosa. I. The isolation of two gastrins from hog antral mucosa. *Gut* **5**:103–114, 1964.
9. Gregory, R. A.: Memorial lecture: The isolation and chemistry of gastrin. *Gastroenterology* **51**:953–959, 1966.
10. Edkins J. S.: On the chemical mechanism of acid secretion. *Proc. Roy. Soc.* **B76**:376, 1905.
11. Cooke, A. R. and Grossman, M. I.: Comparison of stimulants of antral release of gastrin. *Amer. J. Physiol.* **215**:314–317, 1968.
12. McGuigan, J. E. and Trudeau, W. L.: Studies with antibodies to gastrin: Radioimmunoassay in human serum and physiological studies. *Gastroenterology* **58**:139–150, 1970.
13. Trudeau, W. L. and McGuigan, J. E.: Effects of calcium on serum gastrin levels in the Zollinger-Ellison syndrome. *N. Engl. J. Med.* **281**:862–866, 1969.
14. Reeder, D. D., Jackson, B. M., Ban, J. Clendinen, B. G., Davidson, W. G., and Thompson, J. C.: Influence of hypercalcemia on gastric secretion and serum gastrin concentrations in man. *Ann. Surg.* **972**:540–546, 1970.
15. Passaro, E., Basso, N., and Walsh, J. H.: Calcium challenge in the Zollinger-Ellison syndrome. *Surgery* **72**:60–67, 1972.
16. Yalow, R. S. and Berson, S. A.: Radioimmunoassay of gastrin. *Gastroenterology* **58**:1–14, 1970.
17. McGuigan, J. E.: Antibodies to the carboxyl-terminal tetrapeptide of gastrin. *Gastroenterology* **53**:697–705, 1967.
18. Young, J. D., Lazarus, L., Chisholm, D. J., and Byrnes, D.: A radioimmunoassay for gastrin in human serum. (abstract) *Gut* **10**:950, 1969.
19. McGuigan, J. E.: Antibodies to the C-terminal tetrapeptide amide of gastrin: Assessment of antibody binding of cholecystokinin-pancreozymin. *Gastroenterology* **54**:1012–1017, 1968.

REFERENCES

20. Mutt, V. and Jorpes, J. E.: Isolation of aspartyl-phenylalanine amide from cholecystokinin-pancreozymin. *Biochem. Biophys. Res. Commun.* **26**:392–397, 1967.
21. McGuigan, J. E.: Studies of the immunochemical specificity of some antibodies to human gastrin I. *Gastroenterology* **56**:429–438, 1969.
22. Morgan, C. R., Sorenson, R. L., and Lazarow, A.: Studies of an inhibitor of the two antibody immunoassay system. *Diabetes* **13**:1–5, 1964.
23. Morgan, C. R. Sorenson, R. L., and Lazarow, A.: Further studies of an inhibitor of the two antibody immunoassay system. *Diabetes* **13**:579–584, 1964.
24. Jeffcoate, S. L.: Radioimmunoassay of gastrin: Specificity of gastrin antisera. *Scand. J. Gastroenterol.* **4**:457–461, 1969.
25. McGuigan, J. E.: Antibodies to the carboxyl-terminal tetrapeptide amide of gastrin in guinea pigs. *J. Lab. Clin. Med.*, **71**:964–970, 1968.
26. Yalow, R. S. and Berson, S. A.: Size and charge distinctions between endogenous human plasma gastrin in peripheral blood and heptadecapeptide gastrins. *Gastroenterology* **58**:609–615, 1970.
27. Yalow, R. S., and Berson, S. A.: Further studies on the nature of immunoreactive gastrin in human plasma. *Gastroenterology* **60**:203:214, 1971.
28. Berson, S. A. and Yalow, R. S.: Nature of immunoreactive gastrin extracted from tissues of gastrointestinal tract. *Gastroenterology* **60**:215–222, 1971.
29. Yalow, R. S., and Berson, S. A.: And now, "big, big" gastrin. *Biochem. Biophys. Res. Commun.* **48**:391–395, 1972.
30. Walsh, J. H., Debas, H. T., and Grossman, M. I.: Pure nature human big gastrin: Biological activity and half life in dog. *Gastroenterology* **64**:873, 1973.
31. Hansky, J., Soveny, C., and Korman M. G.: What is immunoreactive gastrin? Studies with two antisera. *Gastroenterology* **64**:740, 1973.
32. McGuigan, J. E. and Trudeau, W. L.: Immunochemical measurement of elevated levels of gastrin in the serum of patients with pancreatic tumors of the Zollinger-Ellison variety. *N. Engl. J. Med.* **278**:1308–1313, 1968.
33. Hansky, J. and Cain, M.D.: Radioimmunoassay of gastrin in human serum. *Lancet* **2**:1388–1390, 1969.
34. McGuigan, J. E. and Trudeau, W. L.: Serum gastrin concentrations in pernicious anemia. *N. Engl. J. Med.* **282**:358–361, 1970.
35. Temperley, J. M. and Stagg, B. H.: Bioassay and radioimmunoassay of plasma gastrin in a case of Zollinger-Ellison syndrome. *Scand. J. Gastroenterol.* **6**:735–738, 1971.
36. McGuigan, J. E. and Trudeau, W. L.: Serum gastrin levels before and after vagotomy and pyloroplasty or vagotomy and antrectomy. *N. Engl. J. Med.* **286**:184–188, 1972.
37. Korman, M. G., Hansky, J. Scott, P. R.: Serum gastrin in duodenal ulcer. III. Influence of vagotomy and pylorectomy. *Gut* **13**:39–42, 1972.
38. McGuigan, J. E., and Trudeau, W. L.: Differences in rates of gastrin release in normal individuals and patients with duodenal ulcer disease. *N. Engl. J. Med.* **288**:64–67, 1973.
39. Stern, D. H., and Walsh, J. H.: Gastrin release in postoperative ulcer patients: Evidence for release of duodenal gastrin. *Gastroenterology* **64**:363–369, 1973.
40. Hansky, J., Soveny, C., and Korman, M. G.: Effect of secretin on serum gastrin as measured by immunoassay. *Gastroenterology* **61**:62–68, 1971.
41. Thompson, J. C., Bunchman, H. H., Reeder, D. D.: Effect of secretin on circulating gastrin. *Ann. Surg.* **176**:384, 1972.
42. Isenberg, J. I., Walsh, J. H., Passaro, E., Jr., Moore, E. W., and Grossman, M. I.: Unusual effect of secretin on serum gastrin, serum calcium and gastric acid secretion in a patient with suspected Zollinger-Ellison syndrome. *Gastroenterology* **62**:626, 1972.
43. Bradley, E. L., Galambos, J. T., Lobley, C. R., and Chan, Y.-K.: Secretin-gastrin relationships in Zollinger-Ellison syndrome. *Surgery* **73**:550–556, 1973.
44. Wilhelmj, C. M., McCarthy, H. H., Hill, F. C.: Acid inhibition of the intestinal and intragastric chemical phases of gastric secretion. *Amer. J. Physiol.*, **118**:766–774, 1937.
45. Woodward, E. R., Lyon, E. S., Landor, J., and Dragstedt, L. R.: The physiology of the gastric

antrum: Experimental studies on isolated antrum pouches in dogs. *Gastroenterology* **27**:766–785, 1954.
46. Uvnäs, B.: Discussion of Schofield, B.: Inhibition by acid of gastrin release. In Grossman, M. I. (ed.), *Gastrin*, UCLA Forum in Medical Sciences, No 5, University of California Press, Berkeley, Calif., 1966, p. 186.
47. Polak, J. M., Coulling, I., Doe, W., and Pearse, A. G. E.: The G cells in pernicious anemia. *Gut* **12**:319–323, 1971.
48. Creutzfeldt, W., Arnold, R., Creutzfeldt, C., et al.: Gastrin and G-cells in the antral mucosa of patients with pernicious anaemia, acromegaly and hyperparathyroidism and in a Zollinger-Ellison tumor of the pancreas. *Eur. J. Clin. Invest.* **1**:461–479, 1971.

Computer Applications in Radioimmunoassay

CHAPTER NINE	JOHN U. HIDALGO, M.S.
	MARY IRVIN, M.S.
	LEONARD SPOHRER
	TED BLOCH, M.D.
	HELEN D. BUSBY, R.T., N.M.T.
	MERILYN TRENCHARD, B.S., R.T.
	LINN McMULLIN, B.S., R.T.

In the previous articles, we have had extended discussion of the pitfalls and pratfalls of radioimmunoassay (RIA) techniques. Many of these comments and admonitions are similarly applicable to other "wet" procedures in the nuclear medicine facility. From different perspectives and different procedures, quality control problems ranging from standard curves to interpretation have been discussed.

We have heard the basic tenets of good laboratory medicine reaffirmed, repeatedly:

A. Careful laboratory work is required to produce diagnostic data.
B. Diagnostic data are necessary but not sufficient to establish diagnosis.

These are doctrinaire, but like all virtuous positions they are difficult to achieve and maintain.

Laboratory data reliability (quality control) is a direct result of constant attention to procedural protocol and detail, adequate use of standards, and establishment of an adequate system of checks on all parts of the procedure.

Interlaboratory comparison of test samples is not a substitute for quality control. While these comparisons are useful for regional evaluation of normal values or population comparisons, the task of quality control in a laboratory is a continuous function, as necessary as respiration.

The considerations of good basic technique such as pipetting precision, multiple label reading and the like are well known and may be assumed.

We deal with a systems approach to quality control superimposed on good laboratory technique.

1. The procedure is performed and raw data are recorded by the technologist.

2. The raw data are processed on a desk calculator by the technologist and the results are recorded.

3. These results are delivered to a clerk-typist for transcription to final form and distribution.

There are several places in this sequence where human error can enter, such as the transposition of digits in a numerical report of entering data for computation in incorrect sequence.

We have initiated an automatic data processing system in the Nuclear Medicine Department at Touro Infirmary. The central processing unit (CPU) is an IBM 360-40 equipped with various remote terminals (IBM-1050) on a time-phase system. The remote terminal in the Nuclear Medicine Department is simple to operate. To the technologist, it is an electric typewriter with a few extra keys and the ability to answer back.

To activate the system, the operator depresses the "request" key. The terminal responds by lighting the "proceed" light and turning on the typing module. To direct the attention of the CPU to the stored nuclear medicine programs, the operator then types:

PR SOLVE;

and then proceeds with the entry.

The next item of input is to call up the specific program. This call is a simple three- or four digit alpharmeric code designed for mnemonic value, such as: T-3, for the T-3 test; DGO, for digoxin assay; DGT, for digitoxin assay.

Thus the entry thus far for a T-3 test is:

PR SOLVE; T-3

This is then followed by the patient's demographic data and the input data, such as:

```
                   (1)         (2)     (3)     (4)
PR SOLVE; T-3   xxxxxxxxxx   xxxxx   xxxxx   xxx    (T)
```

where (1) is patient's identification number; (2) is resin count; (3) is total count; (4) is temperature (°C); and (T) is the EOT signal to instruct the CPU that transmission is complete and to proceed with computation.

If this is an outpatient, direct input of demographic data is allowed, for example:

PR SOLVE; T-3 DD JANE DOE, 1212 THIRD STREET, NOLA* DAVIS, 18749 67048 22 (T)

The CPU proceeds by testing input data. For example, in this procedure there must be three pieces of data; the count data must be at least three digits and less than six digits, and the temperature must be two digits unless inputted to the nearest tenth of a degree (as 22.2) in which case three digits are allowed. It further tests that data item (3) is larger than data item (2).

If any data fail the edit test, the system reports the discrepancy and stops. The input must be reinitiated correctly.

If edit detects no errors, the CPU will calculate the ratio, correct for temperature

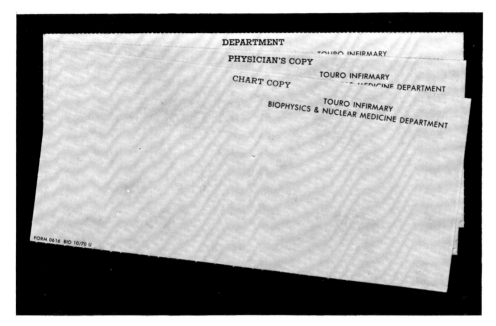

Fig. 1. Report forms used for CPU print-out.

variation, and print the final report on multiple-copy forms (Figure 1) ready for separation and distribution. Before distribution a visual check of input data is performed by the operator technologist.

The report is depicted in Figure 2. The normal values are stored in the program and printed on each report.

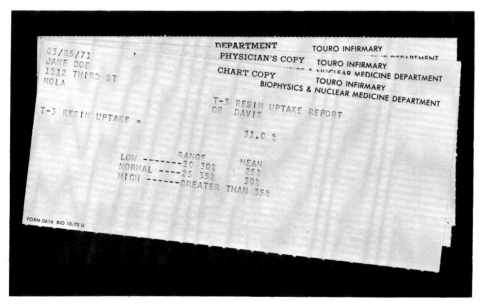

Fig. 2. CPU print-out of final report.

On each day that T-3's are done a test is performed on standard sera of known, nominal value. This is the first entry into the CPU and is done as follows.

$$\text{PR SOLVE; T-3S} \quad \underset{(1)}{\text{xxxxx}} \quad \underset{(2)}{\text{xxxxx}} \quad \underset{(3)}{\text{xxxxx}} \quad \underset{(4)}{\text{xxxxx}} \quad (T)$$

where (1) is resin counts; (2) is total counts; (3) is temperature; and (4) is nominal value.

The CPU performs similar edit checks, and reports derived and nominal values. It computes a correction factor for all subsequent T-3's until a new standard is entered.

With the example of T-3 for experience, let us proceed to similar applications in radioimmunoassay.

As the first example of this group, consider the digoxin assay (DGO). This is not a simple ratio-derived assay, but requires the use of a standard curve. Prior to the entry of the DGO assay data, the standard curve is loaded into the CPU. The procedure for this load routine is

$$\text{PR SOLVE; DGOL} \quad \underset{(1)}{\text{xxxxx}} \quad \underset{(2)}{\text{xxxxx}} \quad \underset{(3)}{\text{xxxxx}} \quad \underset{(4)}{\text{xxxxx}} \quad \underset{(5)}{\text{xxxxx}} \quad \underset{(6)}{\text{xxxxx}}$$

$$\underset{(7)}{\text{xxxxx}} \quad \underset{(8)}{\text{xxxxx}} \quad \underset{(9)}{\text{xxxxx}} \quad (T)$$

where (1) is background counts; (2) through (8) are counts corresponding to trace thru 10.0 ng/ml; and (9) is total counts.

The CPU performs edit checks as before, with the addition of

(a) All counts less than (9)
(b) $2 > 3 > 4 > 5 > 6 > 7 > 8$

If all are correct, the CPU stores this curve until replaced, and reports to the terminal:

DIGOXIN LOAD OKAY 02/27/73

Then we may enter one, or many, DGO procedures as follows:

PR SOLVE; DGO DD SMITH MARY 3355 STONE STREET, NOLA.*,

$$\text{TRENCHARD, 06 22 73} \quad \underset{(1)}{} \quad \underset{(2)}{2045} \quad (T)$$

where (1) is date of sample; and (2) is sample counts.

The CPU establishes a value in nanograms per milliliter from linear interpolation of the standard curve (DGOL) and reports this value as in Figure 3.

If the value lies off either end of the curve, the program will not extrapolate, but rather will report a "less than" or "greater than" value.

The format and technique for a digitoxin assay are so similar that the details are not given here.

Some RIA procedures require a presample regimen which adds a dimension to the automated data processing.

An example of this is found in insulin assay procedures (INS).

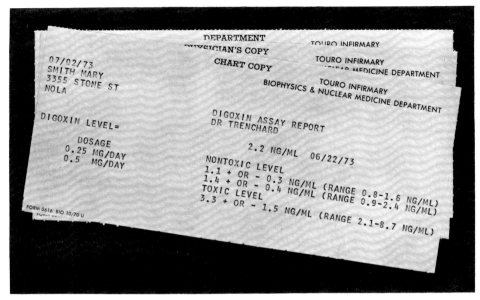

Fig. 3. CPU report of digoxin assay. Value is in nanograms per milliliter from linear interpolation of the standard curve (DGOL).

First, the standard curve is loaded:

	(1)	(2)	(3)	(4)	(5)	(6)
PR SOLVE; INSL	1612	2438	3347	4463	5558	6065

(7)	(8)	(9)	
6663	6714	28191	(T)

where (1) to (7) are sample counts of 320 μv/ml to 3.2 μv/ml; (8) is counts of "zero" sample; and (9) is total counts.

Edit checks are performed. The kit is tested by calculating zero counts as a percentage of total counts. This value must lie between 15 and 30%. The CPU reports back via terminal:

INSL OKAY 02/27/73 CHECK = 23%

With the standard thus loaded, any of several defined regimens may be used.

For a single-sample insulin assay (INSN), the input is:

PR SOLVE; INSN DD SMITH JOHN, 2233 CHARLES STREET, NOLA*,

	(1)	(2)	(3)	(4)	(5)	
SMITH,	02	28	73	5096	0	(T)

where (1), (2), and (3) are date of sample; (4) is sample counts; and (5) is hours fasting.

The report is shown in Figure 4.

There are two stimulation regimens in use, and we describe one of these, the tolbutamide stimulation test (INST).

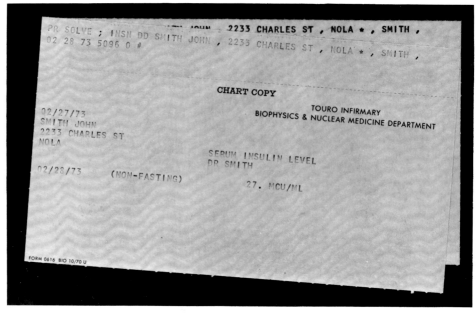

Fig. 4. CPU report of insulin assay.

PR SOLVE; INST DD DOE JANE, 5425 APPLE STREET, NOLA*,

	(1)	(2)	(3)	(4)	(5)	(6)
TRENCHARD, 02	27	73	0	5817	1	2463	2

(7)	(8)	(9)	(10)	(11)	(12)	(13)	(14)	(15)
2771	5	3078	10	3224	20	4002	30	6289

(16)	(17)	
60	5433	(T)

where (1) is test date; (2) is hours fasting (if unknown, O = FASTING); (3) is counts INITIAL SAMPLE; (4) is time to second sample, and (5) is counts in second sample and so on to last sample.

The report is shown in Figure 5.

A similar complexity due to regimen variation is found in the plasma renin (angiotensin) assay (PRA). This results in a series of programs:

> PRAL, standard curve load routine.
> PRA, 3 day 180 mEq sodium diet.
> PRB, 3 day 10 mEq sodium diet, with Hydrodiuril option.
> PRC, random sample, upright, supine, or both.

Similar routines have been established for all the laboratory procedures in the facility.

In the interest of exploiting the advantages of the system, we have also established a batch feature. In this mode the first input is:

PR SOLVE; BATCH (T)

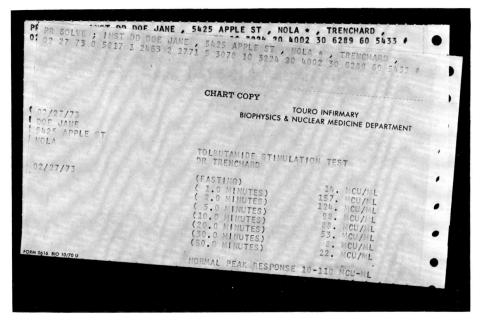

Fig. 5. CPU report of the tolbutamide stimulation test.

and the CPU responds,

BATCH FEATURE STARTED.

In this mode, we then enter procedures as before, and the CPU makes the edit checks and responds with error description or OKAY. We may put in as many procedures as desired, in any mix. Upon completion, we enter:

PR SOLVE; PRINT

and the terminal proceeds to print all the reports.

The system described above offers some advantages for some of the quality control problems of the nuclear medicine facility. There is some simplification of the data handling effort, and the number of human handlings has been reduced. The system is designed to edit input data and to handle procedure control with input standards.

The system can and will undergo further improvement. The use of linear interpolation of the sigmoidal curves for the RIA standards is quite satisfactory, but we feel that there is room for improvement. Linear interpolation places too much emphasis on the validity of individual data points. We have tried all the customary techniques of parametric transformation to linearize the standard curve. We have tried several novel transforms also. In every case, we were able to linearize one part or another of the range. While it is certainly true that the normal measurement is usually the most difficult to verify, we did not feel that the linearization derived was sufficient procedure improvement to justify adoption.

Having gone through our linear phase, we are now looking at other curve-fitting methods. Preliminary results indicate that this area may be more rewarding.

The use of an instructional CPU is analogous to using a cannon to kill a fly. However, there are other advantages in addition to availability.

   a. Monthly or annual summaries are readily available.

   b. Statistical evaluation of normal, abnormal, and population groups are easily done.

   c. If the hospital uses machine billing, this step can be done simultaneously, and reduce paper work in the department.

This system has been in use at Touro Infirmary over 3 years, and we feel that it is effective in achieving the goal of improving quality control in the department. It was designed with the technologist/operator in mind, and has found ready acceptance by these prime users. The system handled over 6000 procedures in 1972, and we encourage the adoption of this type of system for wet laboratory applications.

The Chemical Foundation of Tumor Localization

CHAPTER TEN　　　　　　　　　　　　　　　　　　NED D. HEINDEL, Ph.D.

In the field of nuclear medicine, the discipline of specific organ imaging has been one of the most actively advancing research topics of recent years. This expanding interest derives from the almost immediate clinical application of new scanning diagnostic radiopharmaceuticals in demonstrating anatomic changes in the target organ, and in assisting with the positive early delineation of malignant tumors. The voluminous work in the equally important allied areas of improvement in detection instrumentation, in synthesis and application of ultra-short-lived, cyclotron-produced nuclides such as ^{11}C, ^{13}N, ^{15}O, and ^{18}F, and potential uses of ^{99m}Tc, has caused many researchers to lose sight of the fact that successful organ-imaging measurements are based on biochemical principles that govern the selective distribution of chemical substances in the body. Without a high degree of specificity of uptake for a candidate radiopharmaceutical by a target organ, vis-à-vis the anatomically proximate organs, clear visualization is in many cases an impossibility.

Organ specificity and tumor specificity are two somewhat interrelated concerns of the nuclear medicine physician. It has long been a fond hope of clinicians to have at hand a general radiopharmaceutical that is incorporated in high concentrations in malignancies, such that on scanning the tumor and its metastases are clearly delineated. With the possible exception of the immunological approach using labeled antibodies for tumor specificity, no single underlying *chemical* principle is yet known for general and practical use in tumor localization. Examples of nuclides that concentrate in certain kinds of tumors have been reported, but a common molecular basis for their action does not appear evident.

Organ specificity, however, is a much better understood process, and a limited number of physiological mechanisms has been suggested to justify nuclide uptake by given target organs. Once the radiopharmaceutical gains entry to a specific organ containing a neoplasm, it often experiences differential partitioning between benign and malignant tissue in such a way that tumor detection becomes possible. However, as is all too common with many of the radiopharmaceuticals utilized at present, available tumor incorporation is usually less than that of surrounding healthy tissues, and on imaging the tumor appears as a "cold" area in a high background field. Demarcation of the malignant area is often difficult and uncertain under these circumstances. A "positive" or "hot" uptake is greatly to be preferred.

We attempt to focus on some of the currently established localization mechanisms both of the whole organ and of any potential malignancies that might be present in the organ. In this connection it is important to note that by and large the radiopharmaceutical industry has not yet developed the extensive mass synthesis and screening of candidate agents (some with only very tenuous theoretical basis for expected success) that have characterized the pharmaceutical industry at large, and thus many of the new radiopharmaceutical types tend to extend the few classic mechanisms for localization that have been known for years.

We generally regard the major organ-localizing mechanisms as (1) those involving particulate uptake (including colloidal incorporation, capillary entrapment, and cell sequestration), (2) those involving active transport effects, (3) those involving nuclides that concentrate by dynamic ionic exchange, and (4) those that include species functioning via the antibody–antigen process.

With respect to the tumor-localizing potential of radiopharmaceuticals, definitive chemical mechanisms cannot yet be offered. General explanations invoke (1) an increased permeability of capillaries in tumors for any sort of chemical substance, (2) a higher than normal demand for the components and precursors of cellular metabolism (such as phosphate, nucleotides and nucleosides, peptides, and amino acids), which arises from enhanced tumor growth rate, and (3) application of immune response mechanisms as theoretical foundations for design of new tumor-localizing radiopharmaceuticals.

Microparticulate and colloidal species that localize by phagocytosis or capillary entrapment in liver, spleen, bone marrow, or lung constitute one of the most widely studied families of diagnostic radiopharmaceuticals. The inherent stability of the colloidal state derives from the charged nature of the particle and its ability to affix a counterion surface charge (electrical double-layer effect). The state is therefore maintained by the mutual repulsion of like charges in that the particles do not approach sufficiently closely to coalesce (1). As such, there is virtually an infinite variety of radiopharmaceuticals that might be prepared for imaging the liver, spleen, bone marrow, or lung, since if a particular radionuclide is not itself available as a colloid, an ionic electrolyte form might in all probability be adsorbed onto a nonradioactive colloid such as sulfur or iron hydroxide.

Other than advantages to be gained in detection or in improved radiation dose to the patient by variation in the emitting nuclide component of the colloid, very little difference is to be expected biochemically from one radiopharmaceutical to another. The chemical nature of the tagged atom should not affect the biological properties, since the localization mechanism depends merely on the colloidal or particulate nature of the substance.

Although ^{131}I aggregated albumin is often classed as a nonsoluble particulate substance whose absorption is effected by capillary blockage or by phagocytosis, nuclide attachment appears to be through covalent chemical bonding and not through electrostatic attraction in the double layer. It has been claimed that tyrosine units of the albumin are permanently substituted (presumably at ring positions ortho to the phenolic hydroxyl) by the radioiodine which remains bonded *in vivo* until the material is metabolized by the body (2).

Since phagocytosis of colloids by reticuloendothelial cells, or deposition within

Kupffer cells, reflects a steady-state phenomenon of normal cells (and one denied to cancerous cells), liver and spleen metastases are normally detected as cold areas. An alternative process for splenic uptake is based on the physiological mechanism inherent in the spleen for sequestration of damaged erythrocytes. Johnson (3) has outlined a technique for partial thermal denaturation of red blood cells in the presence of ascorbate-reduced chromium-51 which in its trivalent state is complexed by protein ligands of the erythrocytes. These treated cells are effectively removed in transit through the spleen, and provide one of the better methods for imaging that organ.

Yet another principal mechanism for specific organ localization of radiopharmaceuticals is the one called active transport. This type includes the now classic radioiodide thyroid scanning methods which derive their efficacy from the inner glandular biosynthesis of thyroxine which requires almost 65% by weight incorporation of iodine. In 1966, Bayly (4) decried the lack of application of biochemical research to the development of radiopharmaceuticals based on active transport. In his words, "The great majority of the chemical forms used are either inorganic, labeled drugs or 'tagged' polymers. Only four (cyanocobalamin, selenomethionine, thyroxine and triiodothyronine) can be classified as 'biochemicals.' "

The simple substitution of selenium for sulfur in physiologically active substances (bioisosteric substitution) has been known for some time to produce, in many cases, metabolites of the parent compound (5). For this reason, and because rapid protein biosynthesis of the digestive enzymes occurs in the pancreas, and because ^{75}Se is a convenient gamma-emitting nuclide, [^{75}Se]selenomethionine was a logical candidate for pancreatic visualization, and now represents one of the classic examples of a compound incorporated by an active transport mechanism (6). Spencer (7) has reviewed several claims by previous researchers that [^{75}Se]selenomethionine can concentrate sufficiently in lymphomas to facilitate their detection on external scanning. Furthermore, Britton and Keeling (8) have used this methionine analog for detection of malignant reticuloses. It is probable that what is really being indicated by observation of hot sites of [^{75}Se]selenomethionine incorporation is an abnormally high region of protein biosynthesis which may be suggestive of a malignancy. Several reports have appeared that nonorganic forms of selenium, namely [^{75}Se]sodium selenite and [^{75}Se]sodium selenate have tumor-localizing potential mainly in brain, lung, and skeleton. Cavalieri (9,10) has claimed that selenite (SeO_3^{2-}) as the ^{75}Se isotope is specific for tumors in bone and brain and was not concentrated in normal tissue nor in regions of nonneoplastic disease. Tumor/brain ratios of 13:1 were often obtained in intracranial neoplasms, as well as in tumors of the chest and abdomen. Concentration in primary lung tumors was effected with a 4.6:1 ratio of isotopic tumor uptake compared to plasma. The reports of Spencer (7) on lymphoma uptake are somewhat more cautious. Although he notes that ^{75}Se as selenate or selenite enters mouse lymphomas, the specific activity in the tumor is less favorable than that found for [^{75}Se]selenomethionine. The latter can suitably visualize abdominal lymph nodes, but small lesions may well be missed and false positives may be a problem. All investigators have called for more research to enlarge the data base.

Dynamic exchange diffusion processes permit several metal ions of ionic radius similar to calcium to replace calcium in the bone matrix; radiopharmaceutical design makes use of that principle (e.g., the use of the Sr and Ga nuclides for bone imaging). Recently, Cavaliero (11) has presented data on [^{75}Se]selenite as a tumor-specific agent for bone scanning, which in 22 cases out of 25 was able to pinpoint primary for secondary tumors of bone with a higher degree of accuracy than ^{85}Sr.

Few firm data exist to explain the tumor selectivity of selenium, but some rational hypotheses can be advanced. Selenite is an oxidant, and is reduced *in vivo* by free sulfhydryl moieties whereupon it becomes protein bound. Cavalieri notes the unlikely possibility that intestinal microflora metabolize the reduced selenium into selenomethionine and selenocysteine. Alternatively, McConnell and Roth (12) postulate selenosulfur linkages in the plasma proteins formed directly by the sulfhydryl reduction of the selenite. Tumor concentration may therefore be nothing more than a kinetic effect of the neoplasm absorbing elevated amounts of plasma protein, reflecting its higher rate of protein biosynthesis in the growing cells. Extensive clinical use of ^{75}Se may in all probability be restricted by its long biological $T_{1/2}$ (approximately 65 days) and the subsequent high cumulative total body dose.

On the question of tumor localization mechanisms, especially in the brain, Blau (13) noted that many nuclides and nuclide-carrier combinations of diverse chemical types have been used successfully. Because uptake can be achieved by small ions, proteins, and synthetic high polymers, Blau inferred a nonmetabolic, nonspecific mechanism of concentration in the tumor area related to a breakdown in the blood–brain barrier, with concomitant alterations in the vascular permeability. Bonte et al. (14) have claimed that this same biological hypothesis of altered vascular permeability might be extended to extracranial malignant tumors of soft tissue and bone, and might explain why such a chemically diverse series of nuclide combinations can be absorbed into these neoplasms. This abnormal permeability mechanism for radioactive compound incorporation in tumors (especially in the brain) is based not only on the fact that capillaries in the malignant area are more permeable than those in normal tissue, but is also based on the presence of a greater blood volume and extracellular space in which the radiopharmaceutical can concentrate. ^{99m}Tc pertechnetate has become the preferred brain-scanning agent of recent years, and owes its success at tumor delineation to the abnormal permeability of the malignant mass (15). Other simple inorganic nuclides that have shown remarkable tumor localization effects are salts of ^{197}Hg studied by Sodee (16) and Wolf and Fischer (17) and the ^{67}Ga compounds which are the focus of considerable current research.

Two antitumor compounds presently in the research stage as potential candidates for radiopharmaceutical development are the antibiotic bleomycin and the square-planar metal ion complex, *cis*-diamminedichloroplatinum(II). Bleomycin, isolated from a strain of *Streptomyces verticillus* was described by Fujita and Kimura (18), and its impressive cytotoxicity to dividing cells has been utilized in treatment of certain forms of cancer. A portion of its structure believed to be metal binding is shown in Figure 1; the total structure is not yet known with certainty. Binding of Cu, Co, Zn, Hg, Ni, Fe, Sr, Tc, In, Ga, Au, and Ag takes place with varying stability constants, but preliminary studies have been done only with ^{57}Co, ^{99m}Tc, and ^{111}In (19,20).

Although Renault (21) has shown that this complex antibiotic binds greatly variable amounts of metal ions, the prime consideration in synthesis of a candidate tumor-localizing radiopharmaceutical is not the number of radionuclide atoms bound to each molecule, but rather the nuclear characteristics of the chosen metal

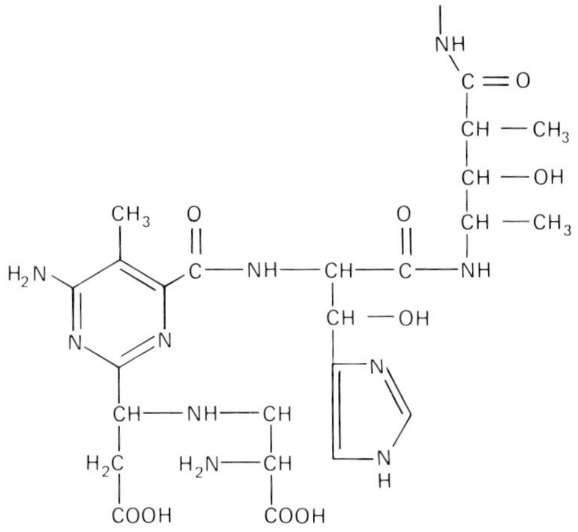

Fig. 1. A portion of the cytotoxic antibiotic bleomycin, believed to be capable of binding radionuclide ions by chelation is shown above. Taken from Renault (21).

ion, whether the binding is sufficiently strong to hold the isotope-carrier combination intact until localization.

It should be noted that in earlier studies Renault (21) had attempted to label bleomycin by electrophilic iodination, and although radioiodine uptake was observed, a rapid *in vivo* dehalogenation caused loss of the labeling before localization occurred. Renault has offered as a reason for this dehalogenation the fact that the polypeptide component of bleomycin carries no aromatic amino acids (such as tyrosine, phenylalanine, or tryptophan) of the type known to bond positive iodine permanently. For this reason, metal binding by chelation offers practically the only way to incorporate a "tagged" atom short of chemically synthesizing the highly complex antibiotic with a ^{11}C, ^{13}N, or ^{15}O in place of one of the stable isotopes.

It is important at this point to consider some of the structural and electronic characteristics of chelates, since considerable present-day research involves the utilization of this method of nuclide attachment to organic carrier moieties. Classically, the search for new diagnostic radiopharmaceuticals focused upon three areas. These were: favorable distribution characteristics of the free nuclide itself; bioisosteric or isotopic synthetic incorporation of a tagged atom within an organic carrier (such as the ^{75}Se for sulfur in methionine, or ^{18}F for the number 5 hydrogen in uracil); and electrostatic adhesion of an ionic nuclide to a colloidal particle. The successful introduction of the technetium-99m chelate of diethylenetriaminopentaacetic acid (DPTA) as a renal scanning agent has greatly spurred research in the chelation method of metal ion binding, and potentially useful radiopharmaceuticals which might be uncovered thereby.

It is not our purpose here to review the field of chelation chemistry either in general or as it applies to medicine, for several excellent texts are available (22,23). The definition of a few terms, however, will be helpful in our discussion. A complex refers to a molecular aggregate of a positively charged metal ion binding nonmetallic donor moieties which can be neutral molecules or anions, but which in every case provide two electrons in the metal–donor bond. The number of species the central metal ion incorporates often exceeds the classic valence of that metal and is called the coordination number. A chelate (from the Greek word *chele,* or claw) is no more than a complex in which two of the donor species are themselves covalently bonded together. The donor groups are more commonly called ligands. A few common examples are the cupric diammine dihalo (e.g., $Cu[NH_3]_2X_2$) and the platinum diammine dihalo complexes (see Figure 2), and the biologically important hemin, cyanocobalamin, and chlorophyll chelates.

The principles of chemistry that dictate what metal ions can be chelated and what types of ligand structures can effect the binding are exceedingly complex and involve many factors, for example, the size and charge of the metal ion, the shape and basicity of the ligating group. Fortunately, the binding potentials of many common ligands for many common metal ions (a quality called the stability constant or K) have been evaluated and tabulated. They are derived from measurement of the equilibrium between the chosen metal ion and its ligand:

$$M + L \rightleftharpoons M - L \qquad K = \frac{[M-L]}{[M][L]}$$

Therefore, the higher the K, the stronger the forces that bind the metal–ligand carrier together (24).

It is not uncommon for some organic ligands, for example, EDTA and DTPA, to affix the metal ion so tightly that the resulting chelate is essentially irreversible. With many chelates, however, a process called ligand–ligand or ligand–metal exchange may take place as represented by the following equations.

$$M + L \rightleftharpoons M - L \begin{array}{c} \nearrow M' - L + M \\ M' \\ \searrow \\ L' \quad M - L' + L \end{array}$$

M and M' = metal ions, M being the radionuclide; L and L' = ligands, L being in the original radiopharmaceutical. These two processes serve to break the original metal–ligand bond, and either release free ligand or free metal ion, or possibly both, into solution.

Renault's (21) studies on the cobalt–bleomycin chelate have shown that at least part of its labeled cobalt can be exchanged onto EDTA if the two components are warmed together:

Co–bleomycin + EDTA $\rightleftharpoons$ Co–EDTA + bleomycin

A group at Hammersmith Hospital, London, (19) has probed the *in vitro* stability of the ^{111}In–bleomycin chelate by incubating samples with metal ions which might

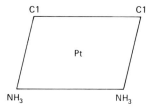

Fig. 2. Structure of the new antitumor platinum complex *cis*-diamminedichloroplatinum (II).

be likely to challenge the chelate *in vivo* (Cu^{2+}, Ca^{2+}, Fe^{2+}, and Fe^{3+}). They have reported that the ^{111}In–bleomycin complex remains chemically stable for up to 60 hours, but they have not, however, studied ligand–ligand exchange *in vitro*. In clinical trials on humans the results were somewhat disappointing. Chelates of divalent radionuclides were almost totally devoid of the normal tumor-localizing activity found in the parent bleomycin, but some complexes of trivalent metal ions (^{67}Ga, ^{111}In, and ^{99m}Tc) did retain most of their biological activity (20). With ^{111}In–bleomycin, which gave the most favorable results, partitioning of the ^{111}In to plasma proteins (chiefly transferrin) began shortly after injection and was nearly complete after 6 hours. Thereafter the distribution and subsequent fate of the nuclide was the same as that observed after injecting ionic indium. Tumor/blood ratios of 30.8:1 were observed with the slow growing RIB.5 tumor and 5.1:1 with the Walker tumor after 72 hours. The amount of activity in the tumor did not alter significantly after 6 hours, and the continued improvement in the tumor/blood ratio merely reflects the progressive reduction in the level of blood activity. At present, it is uncertain whether the degree of tumor localization observed reflects active transport of the bleomycin chelate into the tumor, or whether it represents a concentration of the plasma-bound ^{111}In (19,20). In any event, it does appear that under some conditions the bleomycin–metal ion chelate can dissociate into the nuclide component and free antibiotic.

The same type of ligand–ligand exchange process that may explain the *in vivo* deactivation of bleomycin–metal chelates may be responsible for the potent antitumor activity of diamminedichloroplatinum (II) complexes. Rosenberg and Van Camp (25) have demonstrated that the cis complex shown in Figure 2 causes marked tumor remission in a broad range of approximately 30 animal tumor types. Extensive clinical trials are under way under the auspices of the National Cancer Institute, and definitive reviews of the synthetic chemistry, pharmacology, and structure activity relationship in more than 100 analogs have recently appeared (26, 27). The conclusions to date are that two readily replaceable cis-oriented groups are a requirement for activity (the trans compounds are much less active) (27). Ligand–ligand exchange takes place with close-lying basic receptor sites in DNA to generate inter- and intrastrand chelated cross-linking which interferes with nucleic acid synthesis. Roberts and Pascoe (28) have demonstrated interstrand chelative cross-linking and Harder (29) has postulated an intrastrand connection. Spencer and co-workers (30) have prepared the complex with a ^{193m}Pt nuclide and studied its distribution in mice and rabbits. With the exception of the brain tissue, the labeled complex was widely distributed, with rapid clearance by the kidneys. Wolf et al. (31) evaluated the ^{195m}Pt complex against Walker 256 carcinosarcoma and found that it did not exhibit selective uptake of the antitumor agent, as a tumor/blood ratio of near unity was observed throughout their study. The complex was rapidly cleared by the kidney in a fashion observed for many other chelates.

Both the bleomycin and the platinum complex studies seem to bear out an observation made several years ago by Andrews et al. (32) that "the effect of a drug on a specific organ was often not associated with any unusual concentration there."

Vilhuber (33) has clearly demonstrated that chelates of high stability constant (i.e., those of EDTA and DTPA) show little partitioning or binding to plasma proteins, but localize rapidly in the kidney and are cleared by standard renal processes. Nevertheless, they need not be consigned en masse as renal scanning agents.

Wagner (34,35) has demonstrated their utility for imaging the urinary tract to aid in the diagnosis of obstructive uropathy, and has claimed the superior utility of ^{169}Yb–DTPA as a radiopharmaceutical for cisternography. The chelate diffuses slowly from the cerebrospinal fluid into the vascular system, and then is rapidly cleared by the kidneys.

Tubis (36) prepared chelates of reduced ^{99m}TcO$_4^-$ (which is presumably TcO^{2+} or Tc in the +4 oxidation state) with cystine, methionine, and secretin as potential agents for imaging the pancreas. The latter peptide is well known to trigger fluid and bicarbonate release from the pancreas. Only the ^{99m}Tc–cystine showed much potential for organ imaging, since at short times after injection (ca. 5 minutes), it gave a pancreas/liver ratio of 2:1. Similarly, since tetracyclines have been found to localize in animal tumors, Tubis prepared a ^{51}Cr–tetracycline chelate as a potential tumor-scanning agent. The chelate dissociated *in vivo*, and the distribution observed was the same as that for the liberated ^{51}Cr ion.

Several complexes of ^{51}Cr have, however, been shown to be sufficiently stable for use as tumor-localizing agents. Anghileri (37) has studied the ^{51}Cr complex of glycerophosphate with respect to specificity of uptake into subcutaneously transplanted ependymomas and brain tumors. While most of the complex was cleared rapidly by the kidney and excreted in the urine (59% in 90 minutes) it still concentrated sufficiently well in brain tumors to permit scanning with a tumor/blood ratio of 42:1. Anghileri (37) argues that, based on results of double-labeling, the Cr and the glycerophosphate remain intact until incorporated in the tumor, whereupon the complex is reversed and the glucerophosphate becomes involved in glycolytic pathways. There was also marked incorporation of the chelate into bone, which could be explained as either an active ion exchange of the ^{51}Cr into the bone crystal lattice or incorporation of the phosphate moiety. More recently, the ^{51}Cr complex of alloxantin has been found to localize in five types of animal tumors, including Ehrlich's carcinoma, lymphosarcoma 6C3HED, melanoma B16, lymphatic leukemia BW 5147, and hepatoma BW 7756. Few details have been provided, but again the kidney was a major site of concentration as well as the tumor.

Lastly, rare earth chelates of hydroxyethylenediaminetetracetic acid (HEDTA) were recently reported (38,39) to be useful for skeletal imaging. When complexed to DTPA, the binding is so strong that typical kidney clearance effects are observed. With a much weaker ligand, such as HEDTA, the metal nuclide can dissociate on contact with bone and diffuse into the lattice matrix. Nuclides such as ^{176m}Lu, ^{177}Lu, ^{171}Er, ^{153}Sm, and ^{157}Dy have been employed for this purpose.

The proliferation of chelate-based radiopharmaceuticals in the recent literature is probably a prelude to future rapid developments in this area. One might predict that agents with powerful metal–ligand binding will rapidly accumulate in the kidney, but metal-carrier combinations of weaker binding may partition to organs having either a specific demand for the ion or for its ligand carrier.

In summary, while there is no single underlying chemical foundation for anticipating tumor localization, several principal mechanisms do appear to operate. These include the labeled antibody approach, the abnormal permeability hypothesis, and the kinetic mechanism that relies on enhanced tumor growth rate to absorb, at an elevated rate, the components required for cell reproduction.

REFERENCES

1. Glasstone, S.: *A Textbook of Physical Chemistry*, 2nd ed. D. van Nostrand, Princeton, N.J., 1958, p. 1240.
2. Freeman, L. M. and Blaufox, M. D.: *Physician's Desk Reference for Radiology and Nuclear Medicine*. Medical Economics, Inc., Oradell, N.J., 1971, p. 96.
3. Johnson, P. M.: Clinical scintillation scanning. *The Spleen, Clinical Scintillation Scanning*. In Freeman L. M. and Johnson P. M. (eds.), Paul P. Hoeber, Inc., New York, 1969.
4. Bayly R. J.: A biochemical view of radiopharmaceutical developments. In Andrews, G. A., Kniseley, R. M., and Wagner, H. N., Jr. (eds.), *Radioactive Pharmaceuticals*. U.S. Atomic Energy Commission, Washington, D.C., 1966, p. 53.
5. Burger, A.: *Medicinal Chemistry*, Vol. 1, 3rd ed., Wiley-Interscience, New York, 1970, p. 76.
6. Blau, M. and Bender, M. A.: ^{75}Se-Selenomethionine for visualization of the pancreas by isotope scanning. *Radiology* **78**:974, (1962).
7. Spencer, R. P., Montana, G., Scanlon, G. T., and Evans, O. R.: Uptake of selenomethionine by mouse and human lymphomas with observations on selenite and selenate. *J. Nucl. Med.* **8**:197, 1967.
8. These results have been summarized Taylor, D. M., in *Radioactive Isotopes in the Localisation of Tumours*, William Heinemann Medical Books, Inc., New York, 1969, p. 75.
9. Cavalieri, R. R. and Scott, K. G.: Sodium selenite Se 75. *J. Amer. Med. Assoc.* **206**:591, 1968.
10. Cavalieri, R. R., Scott, K. G. and Sairenji, E.: Selenite-^{75}Se as a tumor localizing agent in man. *J. Nucl. Med.*, **7**:197, 1966.
11. Ray, G., DeGrazia, J., and Cavalieri, R. R.: ^{75}Se selenite as a tumor specific bone scanning agent. *J. Nucl. Med.* **11**:354, 1970.
12. McConnell K. P. and Roth, D. M.: Incorporation of selenium into rat liver Ribosomes. *Arch. Biochem. Biophys.* **117**:366, 1966.
13. Blau M.: In Andrews, G. A., Kniseley R. M., and Wagner, H. N., Jr. (eds.), *Radioactive Pharmaceuticals*, U.S. Atomic Energy Commission, Washington, D.C., 1966, p. 114.
14. Bonte, F. J., Curry T. S., and Oelze, R. E.: Tumor localization by radioisotope scanning. *J. Nucl. Med.* **7**:379, 1966.
15. Bakay, L.: Basic aspects of brain tumor localization by radioactive substances. *J. Neurosurg.* **27**:239, 1967.
16. Sodee, D.: *Radioactive Isotope in Klinik und Forschung*, Band VI. Urban and Schwarzenberg, Munich, 1964, p. 167.
17. Wolf R. and Fischer, J.: Szintigraphische Untersuchungen mit ^{197}HgCl$_2$. In *Proceedings of the 45th Deutsche Roentgenkongress, Wiesbaden*. G. Thieme, Stuttgart, 1965, p. 57.
18. Fujita M. and Kimura, K.: *Progress in Antimicrobial and Anticancer Chemotherapy*, Vol. 2. Tokyo Press, Tokyo, 1970, p. 300.
19. An excellent review of recent bleomycin research is available in Thakur, M. L., Merrick M. V., and Gunasekera, S. W.: Some Pharmacological Aspects of a New Radiopharmaceutical: Indium 111-Bleomycin. In *Proceedings of the IAEA*, SM-171-7, Copenhagen, IAEA, Vienna, 1973.
20. Merrick, M. V., Williams, E. D., Thakur, M. L., Gunasekera S. W., and Lavender, J. P.: ^{111}In-bleomycin: A new tumour Localising Agent, Laboratory and Clinical Experience. Abstracts of the British Nuclear Medicine Society's Annual Meeting, London, April 1973.
21. Renault, H., Henry R., and Rapin, J.: Chelation de cations radioactifs par un polypeptide: La bleomycine. In *Proceedings of the IAEA*, SM-171-24, Copenhagen, IAEA, Vienna, 1973.
22. Murmann, R. K.: *Inorganic Complex Compounds*. Reinhold Publishing Corporation, New York, 1964.
23. Dwyer F. P., and Mellor D. P.: *Chelating Agents and Metal Chelates*. Academic Press, New York, 1964.
24. Sillen L. G. and Martell, A. E.: *Stability Constants of Metal-Ion Complexes*. The Chemical Society, London, 1964.

25. Rosenberg B. and Van Camp, L.: The successful regression of large solid sarcoma 180 tumors by platinum compounds. *Cancer Res.* **30**:1799, 1970.
26. Rosenberg, B. *Anti-tumor platinum compounds. Platinum Met. Rev.,* **15**:42, 1971.
27. Cleare, M. J. and Hoeschele, J. D.: Anti-tumor platinum compounds. *Platinum Met. Rev.* **17**:2, 1973.
28. Roberts, J. J. and Pascoe, J. M.: Cross linking of complementary strands of DNA in mammalian cells by antitumor platinum compounds. *Nature* **235**:282, 1972.
29. Harder, H. C.: Platium coordinated purine dimers. *Biophys. Soc. Abstr.* **11**:299A, 1971.
30. Lange, R. C., Spencer, R. P., and Harder, H. C.: Synthesis and distribution of a radiolabeled antitumor agent: *cis*-Diamminedichloroplatinum(II). *J. Nucl. Med.,* **13**:328, 1972.
31. Wolf, W., Manaka, R. C., and Ingalls, R. B.: Radiopharmaceuticals in clinical pharmacology: ^{195m}Pt-*cis*-diamminedichloroplatinum(II). In *Proceedings of the IAEA,* SM-171-51, Copenhagen, IAEA, Vienna, 1973.
32. Andrews, G. A., Kniseley, R. M., and Wagner, H. N., Jr. (eds.): *Radioactive Pharmaceuticals.* U.S. Atomic Energy Commission, Washington, D.C., 1966, p. 116.
33. Vilhuber, H. G.: Renal excretion mechanisms of radioactive metal chelates. *J. Nucl. Med.,* **9**:357, 1968.
34. Kirchner, P. T., James, A. E., Jr., Reba, R. C., and Wagner, H. N., Jr.: Diagnosis of obstructive uropathy with serial anger camera images. *J. Nucl. Med.,* **12**:444, 1971.
35. DeLand, F. H., James, A. E., Jr., Wagner, H. N., Jr., and Hosain, F.: Cisternography with ^{169}Yb-DTPA. *J. Nucl. Med.* **12**:683, 1971.
36. Tubis, M., Crandall, P. H., Cassen, B., and Blahd, W. H.: Recent developments in new agents, instrumentation and techniques for tumor localization. In McCready, V. R., Taylor, D. M., Trott, N. G., Cameron, C. B., Field, E. O., French, R. J., and Parker, R. P. (eds.), *Radioactive Isotopes in the Localization of Tumours.* Wm. Heinemann Medical Books, Inc., New York, 1969, pp. 87–89.
37. Anghileri, L. J.: Nature and mechanism for concentration of ^{51}Cr complexes in tumors. *J. Nucl. Med.,* **11,** 380, 1970.
38. Yano, Y., Van Dyke, D. C., Verdon, T. A., Jr., and Anger, H. O.: Cyclotron-produced ^{157}Dy—A bone scanning agent, *J. Nucl. Med.,* **12**:407, 1971.
39. O'Mara, R. E., McAfee J. G., and Subramanian, G.: Rare earth nuclides as potential agents for skeletal imaging, *J. Nucl. Med.,* **10**:49, 1969.

The Role of a Radiopharmacist in the Development of a Tumor-Localizing Radiopharmaceutical

CHAPTER ELEVEN   WILLIAM H. BRINER, CAPTAIN, USPHS
(RETIRED)

One should realize at the outset that there is nothing very unique about a tumor-localizing radiopharmaceutical. This is to say that the role of a radiopharmacist in the development of any radioactive pharmaceutical is much the same as that of a product development pharmacist whose practice involves nonradioactive dosage forms. The function in either situation is simply the formulation of an end product (the drug) that is safe and efficacious. The starting material in either case is a chemical compound which someone—a chemist, a physician, or perhaps a chemical engineer—believes to have some merit as a diagnostic or therapeutic agent. In the simplest terms, then, the role of a radiopharmacist is to practice radiopharmacy, a subspeciality of pharmacy concerned with radioactive drugs.

To dispel any notion that radiopharmaceuticals are not drugs in the usual sense of the word, one needs only to examine the appropriate section of a federal law dealing in great detail with this subject. Section 201 (g)(1) of the federal Food, Drug and Cosmetic Act (1) indicates that the term "drug" means, among other things, articles recognized in the official *United States Pharmacopeia* (2), official *National Formulary* (3), or any supplement to either of these; articles intended for use in the diagnosis, cure, mitigation, treatment, or prevention of disease in man or other animals; and articles (other than food) intended to affect the structure or any function of the body of man or other animals.

Another federal law, the Atomic Energy Act of 1954, as amended (4), is also germane to this discussion. This statute assigns certain responsibilities regarding the safe use of by-product material in humans to the U.S. Atomic Energy Commission (AEC). Thus for years it has been a standard condition of licenses issued by the AEC that by-product material shall not be used in humans until its pharmaceutical quality and assay have been established. This same condition is included in licenses issued by the individual states with whom the AEC has entered into agreements delegating the responsibility of medical by-product licensing to these states.

Quite clearly, then, radiopharmaceuticals are drugs (5) and, under the terms of two federal laws, these dosage forms must be treated as such. In addition, laws enacted by state legislatures and delegated to state boards of pharmacy for enforcement are applicable, since all radiopharmaceuticals are prescription drugs under federal law.

Although difficulties may be encountered in the formulation of radiopharmaceuticals intended for oral administration, it should be obvious that formulation misadventures in a drug to be administered by a parenteral route will give rise to much more rapid, and usually more severe, adverse effects in patients who receive these materials (6). Therefore this article considers the development and formulation of radioactive drugs for use by some injectable route of administration.

PRECLINICAL EVALUATION

The professional competence of a radiopharmacist can be quite valuable even before a radioactive drug is ever introduced into human subjects. Federal regulations require that animal tests of a new drug be conducted in order that the safety of the dosage form as finally constituted for use in humans may be evaluated. It is well established that many lower animals tolerate a variety of biological insults, without readily demonstrable adverse effects, to an extent greater than the human body does. With a strong background in the physiological and pharmacological effects of chemical substances, a radiopharmacist is able to avoid many problems in the constitution of a drug dosage form that may not be apparent to a scientist with a lesser pharmaceutical orientation. Since these problems may not be noted when the formulation is administered to the usual laboratory animals, early clinical trials may result in adverse effects in human patients, with possible rejection of a potentially valuable radiopharmaceutical because of these avoidable incidents.

Therefore, as is the case with the development and formulation of any parenteral product, there are several areas of concern with which one must deal if a project is to be carried out in a safe manner. These include:

1. Site and facilities.
2. Personnel qualifications.
3. Establishment of production protocol.
4. Packaging and labeling.
5. Quality control.

SITE AND FACILITIES

In the simplest terms, any area in which radiopharmaceutical formulation work is to be undertaken must be suitable for the task to be accomplished. The ultimate quality of pharmaceutical products is directly related to the environment in which they are produced.

One of the abiding concerns in parenteral radiopharmaceutical formulation is the possibility of contamination of the final product. Contamination may result from microorganisms and metabolic products of their growth, dust, chemicals, other radioactive substances which may be in the vicinity, and so forth. One would not expect a radiopharmaceutical manufacturer to prepare injectable products in a dismal, dusty area devoid of environmental controls. This should be a matter of similar concern in a radiopharmacy located in a hospital or medical center, for a radiopharmacist who undertakes even the simplest of radiopharmaceutical formulation procedures accepts the same responsibility as that assumed by a commercial manufacturer.

A laboratory area in which injectable medications are prepared must be extraordinarily clean. Access to the area must be restricted to those personnel whose duties require their presence. The air supply for a laboratory of this sort is particularly critical. Ideally, air entering a parenteral products formulation area should have passed through a particle retentive filtration system to reduce the ambient dust level to a minimum. The laboratory should be maintained at a slightly higher air pressure than that of surrounding space, to reduce the possibility of the influx of contaminated air from these less critical areas. As an additional precaution, many laboratories utilize strategically located ultraviolet lights of germicidal wavelength to reduce the ambient microbiological background, and require all personnel entering this area to wear sterile caps, gowns, masks, gloves, and shoe covers. At the very least, a sterile enclosure such as a gloved box or a laminar-flow clean bench must be provided to permit aseptic procedures to be carried out in a safe manner.

A radiopharmacist is well versed in the requirements for the site and facilities needed for this type of procedure, and is able to avoid the many pitfalls in the design or modification of space which may be designated for this purpose.

PERSONNEL QUALIFICATIONS

The preparation of parenteral radiopharmaceutical products is a highly complex task which requires highly trained personnel. Because of the limited number of radiopharmacists, this function is frequently delegated to nonpharmacist personnel. It is incumbent upon the nuclear medicine physician who makes this delegation to determine that these personnel have been adequately trained to permit them to perform this task in a fashion that is compatible with patient safety and applicable regulatory requirements. Under federal law (1) all radiopharmaceuticals are prescription drugs (7). Laws enacted by several states require that such medications be compounded and dispensed only by pharmacists or physicians licensed by these states to perform that function. In the absence of a professionally trained and licensed pharmacist, the physician must assume the *total* responsibility for the safety, purity, and efficacy of radiopharmaceuticals prepared in his department, for that responsibility cannot be delegated to unlicensed personnel.

The limited number of radiopharmacists available should not deter nuclear medicine departments from obtaining pharmaceutical advice and consultation when indicated. In many cases, assistance of a strictly pharmaceutical nature may be obtained from a hospital pharmacist at the same institution as the nuclear medicine group that needs this type of assistance. Thus matters relating to environmental conditions that should prevail in any parenteral formulation area, sources of supply of injection vials and their method of preparation and sterilization, additives, filtration and other sterilization methods, and countless other details of drug formulation are well known to hospital pharmacists. This source of information and advice should be utilized by nuclear medicine personnel when pharmaceutical competence is not available within the nuclear medicine group.

ESTABLISHMENT OF PRODUCTION PROTOCOL

Prior to the use of any new drug in human subjects, animal studies are required to demonstrate the safety of the drug. These should be conducted in at least two species, one of which is a nonrodent. It is always advisable to conduct these studies with the same formulation as that expected to be used in humans. Furthermore, it is required by federal authorities that the route of administration in animal studies be that which will be used in human trials (8).

The selection of components and constituents of the product must be given a great deal of attention if pharmacological, physiological, and chemical incompatibilities are to be avoided. Decisions must be made regarding the possible use of additives, such as microbiological preservatives, stabilizers, antioxidants, buffers, and chelating agents, to name but a few. It is quite advantageous to have these problems resolved by one who is competent in the development of an injectable product. A radiopharmacist has this expertise.

Once established, the production protocol must be carried out in every detail in each succeeding batch of product. Even slight variations in preparation methodology can be catastrophic. At the very least, deviation from an established protocol may lead to unpredictable, and at times, undesirable behavior of the product when administered to patients.

PACKAGING AND LABELING

One of the most important of product constituents is the container and closure in which it is packaged. Specifications for injection vials and stoppers must be very carefully established, for both are in intimate contact with the product. They may contribute contaminants or, at times, remove product ingredients from a solution or dispersion by an adsorptive process. The compatibility of a radiopharmaceutical formulation with the container and stopper must frequently be evaluated prior to the final selection of these components. This is particularly important in the case of kits for extemporaneous use in final product formulation, since most of the kit components are not radioactive and may have a prolonged shelf life. These evaluations must be conducted by personnel with a great deal of pharmaceutical orientation in order to avoid serious difficulty with the finished product.

In general, USP Type I glass container (2) should be employed for radiopharmaceuticals. Although this borosilicate glass is quite resistant to chemical attack, it is never wise to reuse these containers. The difficulty in decontaminating and reprocessing injection vials after their use more than outweighs their cost.

One of the most bothersome problems with which a radiopharmacist must deal is that of adsorption of a product constituent, usually the radioactive component, to the wall of the vial (6). Siliconization or other methods designed to block the potential binding sites on the surface of the glass vial must be developed to avoid this phenomenon.

Similar problems must be solved in the choice of a self-sealing closure or stopper for the vial. Closures that are useful for this purpose are formulations that differ widely in their characteristics. They may be composed of natural rubber, synthetic polymers, or, at times, a combination of both. They may also contain a vulcanizing agent, such as sulfur; an accelerator; an activator, such as zinc oxide; fillers; antioxidants; lubricants; and other substances (9). With such diverse composition, the probability of chemical incompatibility of the stopper and the radiopharmaceutical product is significant. Obviously, compatibility studies are always indicated prior to the use of any stopper formulation for a radiopharmaceutical product, and these studies must be conducted by personnel with considerable expertise in the pharmaceutical sciences if difficulties are to be avoided.

QUALITY CONTROL

One of the most important contributions a radiopharmacist can make in the development of a radiopharmaceutical product is in quality control. It is this aspect of nuclear medicine practice that is most frequently poorly understood and ineffectively carried out by nonpharmaceutically oriented personnel. As it applies to radiopharmaceuticals, quality control consists of a series of tests, observations, and analyses which indicate beyond a reasonable doubt the identity, quality, and quantity of all ingredients present in a product (10). In addition, appropriate quality control procedures should demonstrate that the technology used to formulate the drug will produce a dosage form of the highest possible safety and efficacy. It is apparent that quality control is not simply a battery of tests applied to the finished product; rather, it is a concept that must be built into the very design of a drug product and, indeed, embedded in the minds of those involved in the manufacture or production of the medication. There are some general areas of concern in radiopharmeceutical quality control:

1. Selection, storage, and specifications for starting materials.
2. Production protocol.
3. Finished product examination.
4. Recommended storage conditions.
5. Clinical use; materials and methods.

As is apparent, quality control encompasses a broad scope of considerations. In general, however, one may conveniently group these concerns into those that are strictly pharmaceutical in nature, and those that relate to the radiation characteristics of the product.

Pharmaceutical considerations include, among others: (1) appropriate labeling of the product to assure conformance with the requirements of regulatory agencies; (2) gross appearance of the product, including clarity, or freedom from particulate contaminants; (3) if product is a true solution, color, or absence of color, as the nature of the product indicates; (4) particle size distribution analyses, if product is a suspension; (5) pH determination; (6) biological testing, such as safety, pyrogen, and sterility.

Radiation characteristics include determination of radionuclide identity and purity; radiochemical purity; and assay for radioactivity. In addition, cognizance of the relative radiation hazard represented by the product is necessary in the choice of appropriate shielding for the finished product. Not only the shield must be sufficient to provide the required safety to the personnel who will use the product, but the design of the shield must permit withdrawal of the product from the vial without significant danger of contaminating the liquid contents of the vial with microbial agents.

CONCLUSION

The development of a new radiopharmaceutical not only is a complex task, but also involves heavy responsibility relating to the safety of human lives. Although in past years the primary concern manifested by regulatory agencies was directed toward the radiation characteristics of these agents, more recent attention has been directed to the pharmaceutical quality of radioactive drugs. In actuality, the technology of radiopharmaceutical design and formulation includes both aspects. Thus radiopharmacists have an exceedingly important role to play in this research and development area, for their training and experience in both the physical and biological sciences, their expertise in pharmaceutical formulation and quality control techniques, and their knowledge of laws and regulations that apply to all drugs can be invaluable.

REFERENCES

1. Federal Food, Drug, and Cosmetic Act as Amended, United States Code, Title 21.
2. *The Pharmacopeia of the United States of America*, 18th Rev., U.S. Pharmacopeial Convention, Inc., Bethesda, Md., 1970.
3. *The National Formulary*, 13th ed., American Pharmaceutical Association, Washington, D.C., 1970.
4. Atomic Energy Act of 1954 As Amended, United States Code, Title 10.
5. Briner, W. H.: Radiopharmaceuticals are drugs. *Mod. Hosp.*, **95**:110–114, 1960.
6. Briner, W. H.: The preparation of radioactive chemicals for clinical use. *Amer. J. Hosp. Pharm.*, **20**:553–561, 1963.
7. Hauser, J.: U.S. Food and Drug Administration policy on radiopharmaceuticals. In Andrews, G. A., Kniseley, R. M., and Wagner, H. N., Jr, (eds.): *Radioactive Pharmaceuticals*. CONF-651111, U.S. Atomic Energy Commission, Washington, D.C., 1966, pp. 689–695.
8. Anonymous: *Clinical Testing—Synopsis of the New Drug Regulations*. FDA Papers, DHEW, U.S. Food and Drug Administration, Washington, D.C.., March 1967.
9. Avis, K. E.: Parenteral preparations. In Osol, A. and Hoover, J. (eds.), *Remington's Pharmaceutical Sciences XIV*. Mack Publishing Company, Easton, Pa., 1970, pp. 1519–1544.
10. Briner, W. H.: Radiopharmaceuticals. In Sprowls, J. B., Jr. (ed.), *Prescription Pharmacy*, 2nd ed. Philadelphia, J. B. Lippincott Company, 1970, pp. 618–659.

Instrumentation Factors in Visualization of Tumors

CHAPTER TWELVE   C. CRAIG HARRIS, M.S.

INTRODUCTION

The *localization* of tumors by the use of radioactive materials *in vivo* has been an objective of developmental efforts for at least 25 years. The direct *visualization* of tumors by radionuclide imaging is the specific objective that has provided the most impetus to the development of modern radionuclide imaging instrumentation.

The localization and visualization of intracranial neoplasia were among the targets of the very first developments. Moore (1) studied the use of sodium diiodofluorescein labeled with ^{131}I by survey methods involving Geiger-Müller detectors, and later with scintillation detectors. These studies were augmented by those of Ashkenazy (2), and the methods used showed promise. Selverstone (3) used Geiger-Müller detectors and proportional counters in the form of brain probes to detect the uptake of ^{32}P in intracranial neoplasia at the time of surgery. Wrenn et al. investigated with special instrumentation the special properties of annihilation radiation from positron emitters in brain tumor localization (4). With survey methods, only localization was possible; visualization became a reality only with the invention of the "scintillation scanner" (5) and the "scintillation camera" (6). In 1958, Shy et al. (7) published what amounted to a status report on instrumental means of tumor visualization, which indicated advances in development of radiopharmaceuticals and of instrumentation concepts. By this time, radioiodinated human serum albumin labeled with ^{131}I was an agent of choice, having been introduced by Moore (8). Scanning equipment made use of focused collimators (9), large detecting crystals for increased sensitivity, and energy-selective counting (10) to improve visualization by rejection of scattered radiation. Subsequent developments in radiopharmaceuticals and instrumentation have not only improved the visualization of intracranial neoplasia, but they also have made possible the detection and visualization of many other tumors.

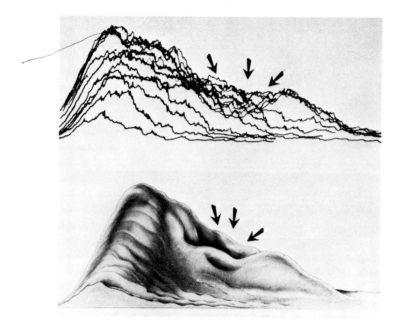

Fig. 1. Three-dimensional presentation of liver radionuclide images. (Top) A composite series of juxtaposed profiles over the liver recorded by strip chart and rate meter at half-inch intervals. (Bottom) Artist's representation of three-dimensional recording. The tumor is visualized as the large solitary depression denoted by arrows. Reprinted from ref. 12.

TUMOR VISUALIZATION BY MEANS OF RADIONUCLIDE IMAGES

This discussion is confined to tumor detection as a function of perception with a two-dimensional image. It is to be noted that image information is really three-dimensional, having two dimensions in space and one in magnitude. Thus three-dimensional forms of visualization are possible (11). Examples of psuedo-three-dimensional images are shown in Figure 1 (12). Additionally, tumors may be visualized by recording the counting rate as a function of time from a selected region of a radionuclide image. Inasmuch as most tumor localization today involves the use of a rectilinear scanner or a scintillation camera for direct visualization by two-dimensional displays, our attention is turned to those concerns, and to the problems peculiar to them.

Visualization of Tumors Depends on Tumor/Nontumor Contrast as Perceived in the Image

The identification of tumors in a radionuclide image depends on the perception of the tumor as an abnormal uptake, or lack of uptake, against a background of normal activity. The perceived contrast, therefore between tumor and nontumor portrayed activities is the basic factor that determines whether or not a tumor will be visualized.

Image tumor/nontumor contrast depends on:

1. Tumor/nontumor count rate ratio.
2. Count density (counts per square centimeter).
3. Display system data manipulation.
4. Display medium characteristics.

The above listing is certainly not all-inclusive, but it does include the principal areas in which instrumental factors make their greatest impact.

It is basic that the tumor/nontumor image contrast depend on the counting rate actually received from the respective regions, which in turn is a function of a considerable number of variables. The perceived contrast also depends, for elementary statistical reasons, on the total number of counts representing the image of the tumor. Inasmuch as the specific area is defined, we speak of the requirement for an adequate count density (see page 109 for a discussion of this term). Manipulation of the image data as a part of presentation by some display system can modify the perceived tumor/nontumor contrast in the image. Present-day imaging is carried out with data processing which varies from essentially nothing to sophisticated data manipulation by computer processing. Finally, the essential characteristics of the display medium itself have a significant material effect on the perceived contrast. One of the advantages of x-ray film as an image medium is its extremely long gray scale. However, Polaroid high-speed film used with scintillation cameras has a notoriously short gray scale. There are other factors involved in tumor visualization, but the above areas constitute those that either can be controlled by the operator or which must be allowed for by the interpreter of the image.

Tumor/nontumor count rate ratio depends on:

1. Tumor/nontumor activity ratio.
2. Tumor location.
3. Collimator properties.
4. Scattering geometry.
5. Spectrometer settings.

Tumor/Nontumor Activity Ratio

The consideration that basically determines tumor/nontumor count rate ratio is the ratio of activities in the tumor and nontumor regions. The lack of specificity of radiopharmaceutical agents for tumor as distinguished from nontumor has been stated. If radiopharmaceutical concentrations in the the tumor are sufficiently high, visualization is relatively easy. If, however, visualization depends on lack of tumor uptake as contrasted by a surrounding region of normal tissue, visualization may become very difficult.

Figure 2 shows a hypothetical field of vision of a detector confined to a 7 cc column of tissue, divided into seven 1 cc blocks. If there is a "normal" distribution consisting of a homogeneous distribution of radioactivity, the contribution of each of the seven 1 cc blocks to the record might be as shown Figure 2a. If a tumor failed to concentrate the radioactive material, and if the tumor were to consist of the central 1 cc block as shown in Figure 2b, the counting rate as perceived from that column of tissue would be diminished by a maximum of 15%. In reality, however, the reduction in counting rate could well be less than 15%, because the

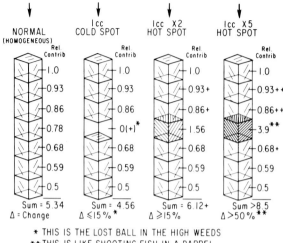

Fig. 2. Theoretical counting rate responses from a detector viewing a 1 cm² column of a homogeneous medium containing radioactivity. The central element of the volume, from 3 to 4 cm deep, is shown with four different loadings of radioactivity: homogeneous (same as surroundings), no activity, twice normal activity, and five times normal activity. Under these circumstances the central volume with no activity and with only twice normal activity may or may not be discernable in a radionuclide image as a lesion.

tumor could scatter radiation from surrounding tissues and thus appear to be less than totally cold. Therefore, looking for a small, cold lesion in a sea of activity is difficult.

If, as shown in Figure 2c, the tumor consists of a hot spot twice as active as its average neighbor of equivalent volume, then the counting rate difference is also 15%, but in the other direction. Thus it is seen that a cold spot is like a "negative times two" hot spot (Figure 2d). Detection of such a hot lesion may be marginal. A volume five times as active as its surroundings is seldom missed. In this situation, its actual size has nothing to do at all with its visualization, only its activity.

Tumor Location

With rectilinear scanners a tumor may or may not be visualized, depending on where it is located, with respect to the focal plane of the focused collimator used to give the scanner directional properties. Figure 3 (13) shows seven images. Each image is that of a cross of blotting paper soaked with cobalt-57. Beginning at the upper left, identical crosses were placed ½ in. from the face of a collimator, and at locations successively 1 in. more distant. The cross that lay in the focal plane is imaged at the top center; this is the only image that is a good representation of the true object. Each of the images contains approximately the same number of counts; the differences between the images lie only in the organization of the given number of counts into a meaningful pattern. From this illustration, we can see two things: (1) If we see something in an image and recognize it by a characteristic shape or size, we may then say that it *was* in the focal plane. (2) If an object lies outside the focal plane, it will not be imaged clearly, but may contribute a significant number of obscuring counts to the image.

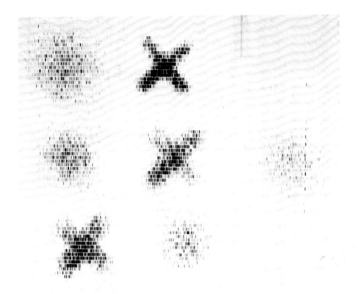

Fig. 3. Images of seven blotting-paper crosses soaked with cobalt-57. Beginning at upper left and proceeding downward, crosses were placed one-half inch from the face of a collimator and at locations successively 1 inch more distance. Distances were, top to bottom, left to right: ½ in., 1½ in., 2½ in., 3½ in. (top of middle column, on focus), 4½ in., 5½ in., and 6½ in. The only image that is a good representation of the true object is the one that did lie in the focal plane. Reprinted by permission from ref. 13.

Scintillation cameras, using parallel-hole collimators, also exhibit a region of best delineation. Figure 4 shows images of a slab phantom imaged with a scintillation camera and a high-resolution camera* The phantom is located at distances beginning with contact with the collimator and increasing 2 in. thereafter. The gradual degradation of the image with increasing object distance is noted.

From these two illustrations, we can draw the conclusion that the rectilinear scanner affords best visualization in the focal plane of the collimator, and that the scintillation camera with a parallel-hole collimator will image most effectively if the target can be placed nearest the collimator. This effect is immediately obvious in any study of the same patient performed on both types of instrument. Therefore actual tumor location, interacting with the basic properties of any collimators used in visualization, is very important in determining the success of visualization.

Scattering

The actual count rate observed from a tumor/nontumor situation, in addition to depending on activity ratios, tumor location, and collimator properties, depends very strongly on how well the tumor can be "seen" when the detector is supposed to be "looking" elsewhere. If gamma rays leave a tumor and proceed elsewhere in the patient and then are scattered into the detector, they will be perceived in the image at the origin of the scattered ray; similarly, a cold lesion will appear to be active if the imaging equipment detects scattered radiation emanating from its volume. Unfortunately, if anything at all happens to photons from the radiopharmaceuticals

---

* Nuclear-Chicago Pho/Gamma HP, with high-resolution collimator.

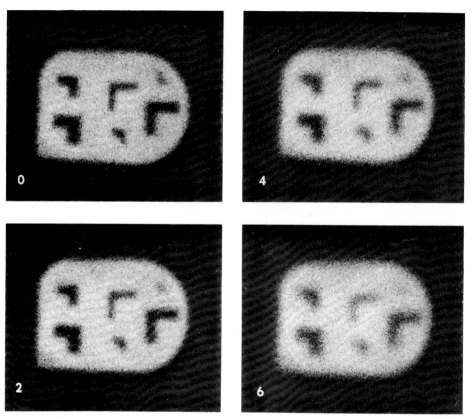

Fig. 4. Scintillation camera images of a slab phantom, homogeneously loaded with ^{99m}Tc, except for fully-displacing voids as shown. The phantom is imaged at four distances with a high-resolution collimator, 0, 2, 4, and 6 in. from the face of the colimator. The gradual degradation of the image with increasing object distance can be seen.

currently used for tumor localization on their attempts to escape a patient's body, there will be scattering rather than absorption. For this reason, the spectrum of radiation escaping from the patient's body, even with a monoenergetic emitter, contains a substantial amount of radiation which has been scattered at least one time.

Figure 4 shows pulse-height spectra observed with a detector fitted with a focused collimator looking at a small source in a low-density background which causes little scattering, and in a tissue-equivalent background in which scattering is substantial. The higher 140 keV peak is associated with the lower 76 keV lead x-ray peak generated in the collimator. Differences in the two spectra are not readily apparent to the untrained eye. For this reason, the spectrum obtained from a scattering medium was elevated by a longer counting time, and is shown in the lower part of the image. The total absorption peak is seen to be broadened on the lower-energy side of the peak; this is a direct portrayal of the detection of scattered photons.

The energy-selective counting modality, made possible by the spectrometric capabilities of sodium iodide crystals, allows us to enhance tumor visualization by appropriate instrument settings. If an acceptance window is chosen to favor the higher-energy regions of the total absorption peak seen in Figure 5, scattered radiation may, to some extent, be rejected. This is illustrated in Figure 6 which shows images of a slab phantom in air (*a* and *b*), and sandwiched between 1½ slabs of paraffin to cause scattering (*c* and *d*). In Figure 6*a* the energy-acceptance band is 135–180 keV; a reasonably crisp image is presented, because scattering takes place only in the source volume itself. Because the slab phantom is imaged in air, an image made with an energy-acceptance span of 100–180 keV, seen in Figure 6*b* is nearly as crisp. With the phantom supported in tissue-equivalent material, however, results are different. In Figure 6*c* a window of 135–180 keV affords a crisp picture, though the effects of some scattering are evident. If the energy-acceptance band is widened to include down to 100 keV, however, there is considerable degradation of the image. An acceptable window for detection of technetium-99m photons has been

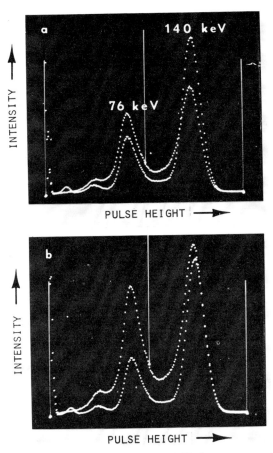

Fig. 5. Pulse-height spectra resulting from ^{99m}Tc 140 keV photons on a sodium iodide scintillation crystal fitted with a focused collimator. (*a*) Spectra are shown for a ^{99m}Tc source in air and in water. The higher total absorption peak at 140 keV is associated with the lower lead x-ray peak at 76 keV, and is for the source in air. Attenuation of the photopeak, increase in scattered radiation, and higher lead x-ray production are seen in the spectrum on the scattering source. (*b*) Similar pulse-height spectra, but counted for a longer time with the source and scattering medium. The total absorption peak is seen to broaden on the lower-energy side, but not the higher-energy limb. The broadening is due to the inclusion of scattered radiation.

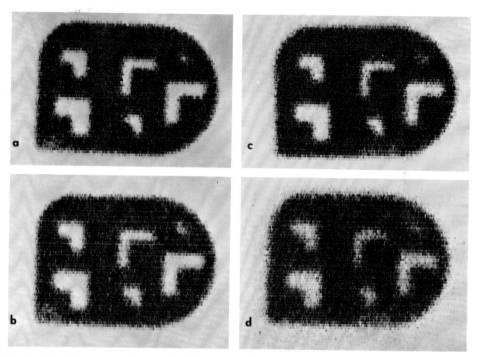

Fig. 6. Rectilinear scanner images of a slab phantom loaded with ^{99m}Tc in air (*a* and *b*) and sandwiched between 1½ in. slabs of paraffin (*c* and *d*). Spectrometer energy-acceptance bands: *a* and *c*, 135–180 keV; *b* and *d*, 100–180 keV. With the phantom in air little difference is seen between the narrow and wide energy bands. In a scattering medium, however, the image is considerably degraded by inclusion of the lower energies (100–135 keV).

derived by several investigators (14,15) and is now in general use. It is especially distressing to see radionuclide images in the literature and in commercial advertising in which this important principle has not been properly embraced.

The principle of offsetting the energy-acceptance window is easily accomplished with rectilinear scanners with no ill effects except perhaps that of lowering the observed counting rate. A similar process has been proposed for scintillation cameras (16). Investigation reveals that the result may well be dependent on the properties of individual instruments. For example, an experiment similar to that shown in Figure 6 was carried out on our scintillation camera,* and the results are shown in Figure 7. In Figure 7*a* the slab phantom was imaged with a "15% window" while supported between 1½ in. thick slabs of paraffin. In Figure 7*b* the image was obtained with a 20% window. Examples of an asymmetrical window are shown in Figures 7*c* and *d*. In Figure 7*c* the window was first centered as a 10% window, and then opened to 20%; the increase in window width was on the high-energy side only. In Figure 7*d* a 5% window was centered and then opened to 20%. It is quite evident that the sharpest image is shown in Figure 7*d*, with a markedly detuned window. It is also evident, however, that the image compared to that in Figure 7*c* is decidedly nonuniform in areas that should portray a uniform distribution of activity. Thus the increased sharpness may come at the expense of nonuniformity. Figure 8 (17) shows

* Nuclear-Chicago Pho/Gamma HP.

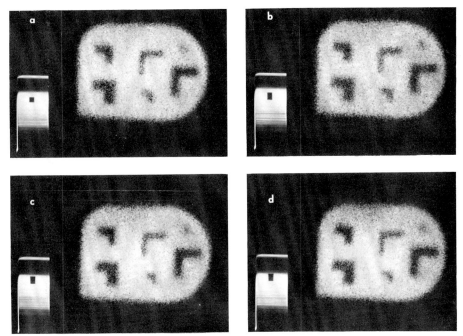

Fig. 7. Scintillation camera images of slab phantom sandwiched between 1½ in. slabs of paraffin. Energy acceptance bands: *a*, 15% window, centered (130–150 keV); *b*, 20% window, centered (126–154 keV); *c*, 10% window, centered, then opened to 20%; *d*, 5% window, centered, then opened to 20%. The image at *d* is perhaps the sharpest, but shows considerable nonuniformity of a homogeneous distribution (hot spot in center).

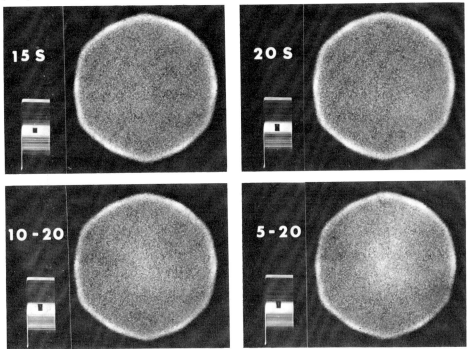

Fig. 8. Flood pictures (collimator off) under the same conditions of Figure 7 show the field nonuniformity. If the scintillation camera is to be used with an off-set window the phototubes must be retuned for uniformity. Reprinted by permission from ref. 17.

flood pictures from the same instrument with the windows the same as in Figure 7. The field nonuniformity is evident. The lesson to be learned here is an obvious one. Detuning the window with a scintillation camera may well enhance the quality of the image in terms of delineation of detail. It might, however, lead to unwholesome and perhaps even dangerous nonuniformities in the image unless the phototubes are adjusted to achieve uniformity. Before any studies are performed on patients with an offset window, the user of each individual camera thus involved should perform the test illustrated in Figure 8, and retune the phototubes for uniformity of field (16).

Spectrometer "Calibration"

Spectrometer settings therefore can be extremely important in the control of scattered radiation and its influence on tumor visualization. There is an additional aspect in which spectrometer settings can be equally important, particularly if the calibration or "peaking," of a spectrometer is carelessly done. Careless calibration of the energy scale, compounded by use of an excessively active source (such as the patient himself or the loaded syringe intended for him) can produce unexpectedly bad results. Figure 9 (17) shows at the top a typical spectrum obtained when a

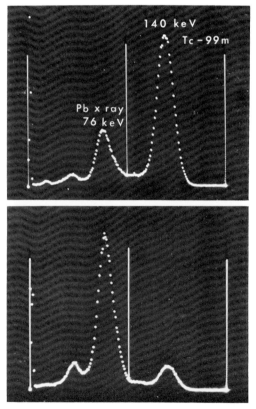

Fig. 9. Pulse-height spectra, ^{99m}Tc on NaI (Tl). (Top) Pulse-height spectrum from weak source near focal plane of focused collimator. (Bottom) pulse-height spectrum from extremely strong distributed source. The total absorption peak at the bottom is actually higher in content than in the lower figure, but is dwarfed by the extremely high lead x-ray peak at 76 keV. Reprinted by permission from ref. 17.

Fig. 10. Images of slab phantom with spectrometer tuned to 140 keV photons (bottom), and with lead x rays (76 keV) boosted to 140 keV equivalent (top). This kind of "mispeaking" of a spectrometer can lead to totally false results.

properly weak source is presented to a detector fitted with a focused collimator. The spectral peak identified as the lead x ray at 76 keV, shown also in Figure 5, is the result of the stopping of 140 keV photons in the lead collimator close to the detector. When an intense source of technetium-99m is used for energy calibration, a collimator stops many more photons than it passes; hences the lead x-ray peak can completely dominate the 140 keV desired peak. If the lead x-ray peak is mistaken for a misplaced 140 keV peak and boosted to a 140 keV equivalent, spectacularly bad scans can result. Figure 10 shows at the bottom an image of a slab phantom with the spectrometer properly calibrated. At the top is shown an image of the same phantom, but with the lead x ray boosted in pulse height to the equivalent of 140 keV. This problem has been known to occur in actual clinical procedures. It need not occur if careful and proper methods are used along with sufficiently weak sources.

Count Density—Statistical Considerations

Tumor perception in radionuclide images, particularly when detectability is marginal, depends on the ability of the eye to distinguish the tumor from its surroundings. This interpretation is subject to the same considerations of statistical expectations to which any other estimate of radioactivity must adhere. If several images of the same target are made, they may or may not look like each other. Indeed, the number of counts acquired from a specific portion of the subject may be described by exactly the same statistical concerns as the same number of counts obtained on samples in a well counter. This basic fact, coupled with the properties of display systems and recording media used today, place a requirement that sufficient count density be embraced as a basic standard of technique.

Count density (counts per square centimeter)* is related to the information content of the image and is (for scanners) related to other parameters:

$$\frac{\text{counts}}{\text{cm}^2} = \frac{\text{counts}}{\text{minute}} \times \frac{1}{\text{speed (cm/minute)}} \times \frac{1}{\text{spacing (cm)}}$$

Counts per minute are determined by prospective survey of the patient; spacing is usually arbitrarily prechosen. The required speed to achieve a desired count density may be solved for explicitly; alternatively, a fixed speed may be used and the resulting count density determined.

Reproducibility of technique requires adoption of some desired count density at a specified anatomic location; good visualization of tumors requires that the desired count density be high enough for a statistically sound estimate of tumor/nontumor contrast.

Figure 11 (13) shows four images of a phantom which simulates a lateral brain scan. The preliminary setup spot is at the greatest activity over the sagittal sinus.

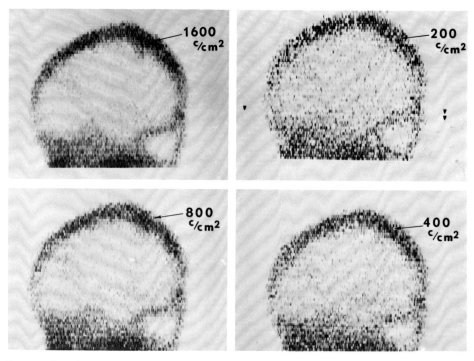

Fig. 11. Four images of a phantom which simulates a lateral brain scan, at four different count densities at the setup point (superior saggital sinus). A mock lesion is clearly seen anteriorly in the high-count-density image. The lesion is not nearly so evident at lower count densities. Reprinted by permission from ref. 13.

---

* The use of "information density" or "I.D." expressed in counts per square centimeter is an improper use of a specific term from another field of science. The misuse can perhaps be condoned, however, if it will cause improvement in imaging technique by use of higher information density.

Count densities at this point are noted in the illustration. If only the image with the lowest count density is considered, one sees a ragged, but apparently normal, distribution of activity. When the image with the highest count concentration is examined, however, a small region of increased activity is seen anteriorly. This increase in activity is real, for it is an additional radioactive source added to the phantom. Once visualized in the high-count-density image, this lesion may be seen in the other images, but perhaps would never have been seen in the lowest-density image.

This principle may be extended to images obtained with the scintillation camera. Figure 12 shows images of the same phantom, but with a different "tumor" and containing different count concentrations. In Figure 12a there are about 2000 counts/cm² at the setup spot, or a total of 500,000 counts. The images progress downward in counting rate by factors of 2, from 500,000 counts in Figure 12a to 16,000 counts in Figure 12f. This affords a dramatic portrayal of the "disappearance of a tumor" when count densities become inadequate for mental computation of tumor/nontumor contrast.

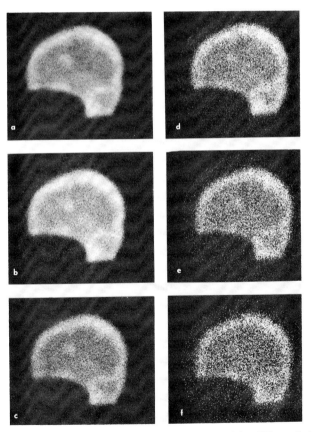

Fig. 12. Six scintillation camera images of the same phantom, but with a different lesion, imaged with decreasing number of total counts. Beginning at (a), the image contains 500,000 counts. Each successive image has half as many counts, with 16,000 counts constituting the image at (f). The tumor is seen slowly to disappear with decreasing total count.

Fig. 13. Images of the brain phantom with a barely detectable lesion, imaged with 2 mm spacing (top) and 3 mm spacing (bottom). The tumor is more evident in the image with finer spacing. Not all the visualization of the tumor is due to the darker scan.

With rectilinear scanners, even when the total number of counts in the image is held constant, differences in other parameters may affect tumor visualization. Figure 13 shows two views of the same brain phantom obtained with a marginally detectable tumor placed anteriorly, supratemporally. The upper image was obtained with 2 mm spacings, and the lower with 3 mm spacing. The total number of counts in both images is approximately the same. In the upper image, however, the smaller spacing affords more opportunities to "write down the answer," and makes the tumor definitely more evident. (The upper image is darker, but not all of the effect is due to the extra blackness.) We are still investigating this particular phenomenon; to date, it is reliably reproducible. Scan spacing can then be seen to be a significant factor in tumor visualization, and should be approximately small. The scan spacing of one-eighth to one-tenth of the diameter of the resolution circle of the collimator seems, in our investigations, to be near optimum.

Display System Data Manipulation

Tumor visualization in a radionuclide image can be affected seriously by modifications of the data in display systems. Some of these modifications are either inherent or accidental; others are deliberate attempts to increase the information evident in the image.

Simple and Inherent Forms of Manipulation

Even where "one-mark-for-one-count," direct photorecording is employed in rectilinear scanning, there is modification of apparent information (compared with, say, dot recording on Teledeltos), because of the long contrast scale available in recording on x-ray film. This represents an inherent influence on the image. However, background erase and contrast enhancement cause specific modifications of the image contrast, and usually are employed for definite purposes. Such must be used with care, however, where the only image obtained is subjected to these modifications, for information removed cannot be restored.

A simple form of data manipulation is available, for deliberate or unwitting use, when scintillation camera imaging with a cathode ray tube display is employed. First, if "preset count" is used to govern the process, acquisition of counts from clinically irrelevant anatomy dilutes the statistical validity of information from the region of interest. Second (and *not* directly related to the first item), if the display dots are poorly or nonuniformly focused on the face of the cathode ray tube, modification of the image can result. These effects may be seen in Figure 14, which shows a phantom imaged with 250,000 counts. In the lower left image, the 250,000 counts are obtained from the region of interest and the visualization of the tumor is good. In the upper left image, count density in the region of interest is diminished because of the inclusion of count from regions not of interest, and tumor/nontumor contrast is reduced.

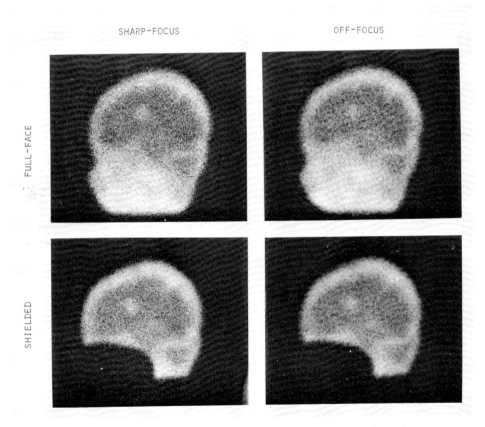

Fig. 14. Scintillation counter images, analog mode, of lateral brain phantom. When using preset count imaging, and when counts are recorded from clinically irrelevant areas, the count density in the region of interest is reduced. This in turn hampers tumor visualization. Additionally, the focusing of dots on the cathode ray tube face also seems to diminish tumor visualization.

Figure 14 also shows that, with the presentation of analog data with this instrument, tumor/nontumor contrast is lowered only if the individual dots presented by the cathode ray tube display are not carefully and uniformly focused. This can occur when the intensity level is changed without a concurrent readjustment of focus and astigmatism. The defocusing presented in Figure 13 was done to achieve some data blending, and required the use of a defocused, but uniform, dot. Nonuniform dot defocusing can cause serious loss of information.

Interestingly enough, when the experiment of Figure 14 was repeated by playing back images of the phantom from videotape* storage in digital form, the lesson of Figure 14 did not apply. While masking of the subject to require that counts come from a specific region of interest still results in better visualization of the tumor, defocusing of the dots in the matrix display actually improves tumor visualization and even achieves a sort of contrast enhancement (see Figure 15).

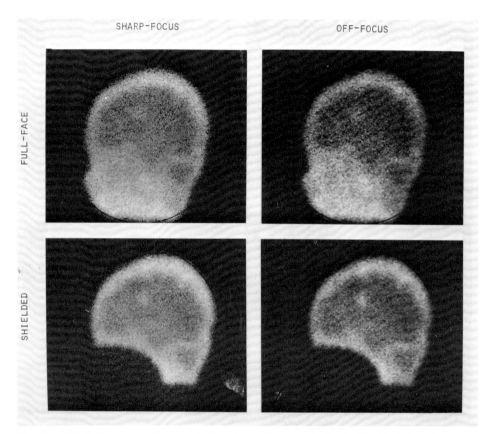

Fig. 15. Scintillation camera images of lateral brain phantom, similar to Fig. 14, but made on playback of digital data from videotape. Shielding uninteresting facial activity seems to improve tumor visualization, as in Fig. 14. The focused dots, however, seem to improve tumor visualization in this form of display.

---

* Nuclear-Chicago data store/playback accessory.

Simple Computer Processing

When image data are recorded in some form of storage that allows an image to be generated from them without destruction or modification of the original data, extensive data manipulation may be attempted to improve tumor visualization. A complete treatment of these methods is outside the scope of this discussion, and is extensively covered in the literature of the subject. A brief treatment, however, may serve to illustrate the value of such methods in tumor visualization.

When stored for computer processing, image data usually are digitized in both magnitude and position, with the result that they consist of a certain number of counts stored in a picture element. For each picture element to contain a statistical number of counts, the number of picture elements is limited. When an image is formed from an untreated picture element format, an unaesthetic image is often the result. Such images can be quite crude, hampering any attempts at tumor visualization. A relatively simple computer operation (18) performs an interpolation between picture elements, and displays the synthesized information in a form much easier to see. Figure 16 is reprinted from reference 18 and illustrates the interpolative display method.

Smoothing and filtering are relatively simple methods of manipulating image data to make more evident the information content of the image. The simplest method of filtering image data is that of spatial averaging, first used to eliminate excessive high-frequency variations in the image (19,20). Spatial averaging involves the replacing of the content of a picture element with the average of that element and those surrounding it. In the simplest form, the weights in the spatial dimensions are equal, and all elements involved have an equal effect in the filtering. Such filtering actually removes information, but often the information that remains is more evident to the eye. There is little evidence in the literature that simple data averaging is currently in use.

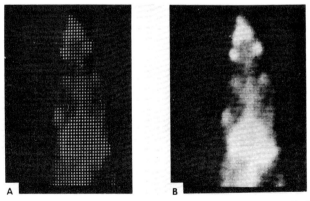

Fig. 16. Computer-processed images showing the effect of interpolation during display. The image in (a) contains 4071 displayed points without interpolation, while that in (b), using the same data in core, contains 16,146 displayed points. The subject imaged is a 280 g rat with a 0.28 mCi dose of ^{67}Ga scanned after 24 hours. Despite the fact that (b) contains no more information than (a), it is much easier to interpret. From ref. 18.

A slightly more complicated form of digital filtering, using weighted averages, in various formats, of the content of elements surrounding the given element can also be used to enhance image data. For example, a weighted averaging using the point-source response of a collimator as the weighting function was used in early developments (21) and has been used since. Figure 17 shows an example of a gallium-67 scan unfiltered (*A*), and the same data displayed with a 7 × 7 element weighted averager (*B*). Such methods seem definitely able to enhance tumor visualization, particularly when count density is low. Simple weighted averaging, combined with interpolative display, requires only a simple computational system and may well be worth the modest investment in terms of improved tumor visualization (18).

More Sophisticated Computer Processing

Where more computational power is available, more complicated forms of data manipulation may be undertaken. Rather than fixed averaging or filtering as mentioned above, one may undertake dynamic or variable filtering or averaging. At low

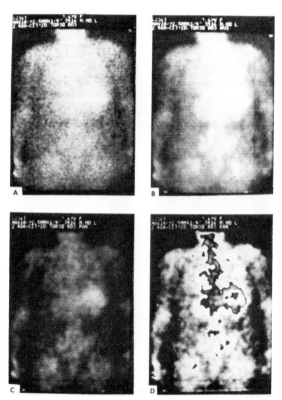

Fig. 17. An example of the correction for scatter and septum penetration. The images are from a Ga-67 scan of a 67 year old female having pseudolymphoma. The raw data are shown in (*A*). (*B*) is the same data averaged with a 7 × 7 element weighted averager. Note the improved smoothness of the data, but also the generally hazy background throughout the image. (*C*) is the same data as (*B*), corrected for scatter and septum penetration. Note the decrease in the hazy background due to scatter and septum penetration, the elimination of counts outside the body area, and the higher peaks and lower valleys. (*D*) is the multicycle contrast enhancement display of the data in (*C*). From ref. 18.

count densities, statistical variations are large; removal of high spatial frequencies is necessary for good visualization, but comes at the expense of contrast in the displayed image. As count densities increase, statistical variations decrease, lessening the need for removal of high-frequency components. Varying the magnitude of the filter with count density is known as dynamic, or variable, filtering. Dynamic filtering and averaging have been shown to be of value in tumor visualization (22).

These methods of resolution recovery can be augmented by still more sophisticated techniques such as "dot shifting." In this process a given count is shifted, in the display, from its actual recorded location in space to a location of more probable origin, based on the distribution of counts in its neighborhood (22). A combination of dot shifting in connection with variable filtering can produce striking results in enhancement of tumor visualization.

Computer processing of image data can assist greatly in removing from the image the deleterious effects of scattering of photons emerging from the patient. We have seen before (Figure 5) that spectrometer settings alone cannot adequately reject scattered radiation. A processing method has been developed to correct for scatter and septum penetration (18), and a result is shown in Figure 17c. The distributions of activity in the gallium-67 scan are considerably more evident than in the original data.

Figure 17 also shows a multicycle contrast enhancement display of the same data. This type of display enhances tumor visualization considerably.

Yet another form of manipulation of data to enhance tumor visualization is possible with a program that corrects for organ motion caused by breathing (23). Data are stored in a series of short images, the centroids of which are located by data analysis. A composite image, with centroids aligned and superimposed, results in a nearly motion-free image.

Many methods, both simple and sophisticated, are available to enhance tumor visualization through the use of data processing equipment. Such equipment is, however, more expensive than the basic imaging equipment alone. Whether the cost can be borne in routine studies, or whether it must be restricted to special studies, remains to be seen.

Display Medium Characteristics

The display medium on which the image is offered can seriously affect the tumor/nontumor contrast in several ways. First, nonlinearities and nonuniformities in the medium can suppress or distort information. The usual form of rectilinear scan imaging by use of exposed dots on x-ray film has at least the virtue of a very long gray scale. If statistical requirements can be met, this can be a very satisfactory medium for tumor visualization. The same generally applies to recording of scintillation camera images on transparency film (35 or 70 mm). However, the use of high-speed Polaroid film for recording scintillation camera images results in images of extremely short tonal range. The result is data compression which can seriously suppress information (24).

The use of color in image recording provides a basis for data expansion, and has been used even in connection with computer processing. This form of image display, making use of the special properties of a special medium, may well increase in the future. It remains to be seen, however, whether the improved tumor visualization offered by color is sufficient to justify the increased cost and difficulty in filing and dissemination of procedure results.

SUMMARY

It can be seen from the foregoing information that a study of *all* the pertinent instrumentation factors in visualization of tumors would result in a very extensive treatise indeed. This discussion has been an attempt to present the salient features, some of which are old and obvious, and some of which are only now becoming well understood.

Recognition of tumors in a radionuclide image is dependent upon recognition of the abnormal as distinguished from the normal. Use of techniques that lead to reliable and reproducible imaging results are most valuable for reliable tumor visualization. Where the additional cost can be supported, computer processing of image data can significantly enhance the visualization process.

REFERENCES

1. Moore, G. E.: Use of radioactive di-iodofluorescein in the diagnosis and localization of brain tumors. *Science* **107**:569, 1948.
2. Ashkenazy, M., Davis, L., and Martin, J.: Evaluation of technic and results of radioactive diiodofluorescein test for localization of intracranial lesions. *J. Neurosurg.* **8**:300, 1951.
3. Selverstone, B., Sweet, W. H., and Robinson, C. F.: Clinical use of radioactive phosphorus in the surgery of brain tumors. *Ann. Surg.* **130**:643, 1949.
4. Wrenn, F. R., Jr., Good, M. L., and Handler, P.: The use of positron-emitting radioisotopes for the localization of brain tumors. *Science* **113**:525, 1951.
5. Cassen, B., Curtis, L., Reed, C., and Libby, R.: Instrumentation for I^{131} use in medical studies. *Nucleonics* **9**(2):46, 1951.
6. Anger, H. O., Mortimer, R. K., and Tobias, C. A.: Visualization of gamma-ray emitting isotopes in the human body. *Proc. Int. Conf. Peaceful Uses At. Energy,* **14**:204, 1956.
7. Shy, G. M., Bradley, R. B., and Matthews, W. B., Jr.: *External Collimation Detection of Intracranial Neoplasia with Unstable-Nuclides.* E. & S. Livingstone, Ltd., Edinburgh, 1958.
8. Moore, G. E.: *Diagnosis and Localization of Brain Tumors: A Clinical and Experimental Study Employing Fluorescent and Radioactive Tracer Methods.* Charles C. Thomas, Springfield, Ill., 1953.
9. Newell, R. R., Saunders, W., and Miller, E.: Multichannel collimators for scanning with scintillation counters. *Nucleonics* **10**(7):36, 1952.
10. Allen, H. C., Jr., Risser, J. R., and Greene, J. A.: Improvements in outlining of thyroid and localization of brain tumors in the application of sodium iodide gamma-ray spectrometry techniques. In *Proceedings of the 2nd Oxford Radioisotope Conference*, Oxford, 19–23 July 1954, Vol. 1. Butterworth Press, London.
11. Tauxe, W. N.: Non-analog displays for quantitative visualization. In Kenny, P. J., and Smith, E. M. (eds.), *Quantitative Organ Visualization in Nuclear Medicine.* University of Miami Press, Miami, Fla., 1971, p. 695.
12. MacIntyre, W. J. and Christie, J. H.: The scanning system and its parts. In Kniseley, R. M., Andrews, G. A., Harris, C. C. (eds.), *Progress in Medical Radioisotope Scanning.* TID 7673, U.S. Atomic Energy Commission, 1963, p. 3.

REFERENCES

13. Harris, C. C.: Image production with rectilinear scanners. In Gottschalk, A. and Potchen, E. J. (eds.), *Diagnostic Nuclear Medicine (Golden's Diagnostic Radiology Series.* Williams & Wilkins, Baltimore, in press.
14. Harris, C. C.: Certain fundamental problems in scanning. In Quinn, J. L., III (ed.) *Scintillation Scanning in Clinical Medicine*, W. B. Saunders Company, Philadelphia, 1964, p. 1.
15. Rollo, F. D. and Schulz, A. G.: Effect of pulse-height selection in lesion detection performance. *J. Nucl. Med.* **12**:690, 1971.
16. Sanders, T. P., Sanders, T. D., and Kuhl, D. E.: Optimizing the window of an Anger camera for ^{99m}Tc. *J. Nucl. Med.* **12**:703, 1971.
17. Harris, C. C.: The care and feeding of medical nuclear instrumentation. *Sem. Nucl. Med.* **3**(3):225-238, 1973.
18. Bell, P. R., McClain, W. J., Ross, B. A., McKeighen, R. E., East, J. K., Moore, E. C., Edwards, C. L., and Goswitz, F. A.: A Computer-Based Scanning System. Technical Manual ORNL-TM-4038, Oak Ridge National Laboratories, January 1973.
19. Brown, D. W.: Digital computer analysis and display of radioisotope scans. *J. Nucl. Med.* **5**:803, 1964.
20. Kuhl, D. E. and Edwards, R. O.: Perforated tape recorder for digital data store with gray shade and numeric output. *J. Nucl. Med.* **7**:272, 1966.
21. Sprau, A. C., et al.: A computerized radioisotope scan data filter based on system response to a point source *Mayo Clinic Proc.* **41**:585 (1966).
22. Pizer, S. M.: Digital spatial filtering and its variations. In Kenny, P. J. and Smith, E. M. (eds.) *Quantitative Organ Visualization in Nuclear Medicine.* University of Miami Press, Miami, Fla. 1971, p. 581.
23. Oppenheim, B. E.: A method using a digital computer for reducing respiratory artifact on liver scans made with a camera. *J. Nucl. Med.* **12**:625, 1971.
24. Harris, C. C., and Goodrich, J. K.: Imaging devices: Comparative physical properties. In Gilson, A. J. and Smoak, W. M., III (eds.), Springfield, Ill., 1970, p. 135.

Development and Utilization of Tumor Localizing Radiopharmaceuticals (^{111}In and ^{67}Ga)

CHAPTER THIRTEEN MORTON B. WEINSTEIN, M.D.

INTRODUCTION

Cancer comes in a large variety of sizes and shapes, and it would be difficult if not impossible to identify a single characteristic that is common to all neoplastic diseases. It is therefore highly unlikely that a single diagnostic procedure using radiotracers or other modalities would be applicable in the diagnosis of all cancers. The failure to identify a single unique feature that is characteristic of cancer has resulted in the development of imaging procedures that are of a low order of specificity but are reasonable sensitive. The noninvasive techniques of nuclear medicine allow procedures to be performed without patient discomfort. There are many standard imaging procedures that detect pathophysiological changes in the *region* of a cancer as opposed to in the cancer per se. Bone scanning with ^{99m}Tc polyphosphate is an excellent example of this. We are able to detect metastatic breast carcinoma to bone before there is any evidence of change in the routine roentgenographic bone survey (including retrospective examination of the films). Brain scanning, liver scanning, and pancreatic scanning are three additional areas that have been extensively evaluated with a variety of radiopharmaceuticals and instruments to detect the presence of neoplastic disease. Table 1 summarizes some of the procedures that are used in our nuclear medicine laboratories for the detection of neoplastic disease. The mechanism whereby the neoplastic lesion is defined has been described. The mechanism whereby indium and gallium concentrate in tumors is currently unknown. Initial observations were serendipitous, but current investigations (1) have demonstrated concentration of these trivalent cations in the lysosomal fraction of cells (2). Why tumor cell lysosomes have an avidity for gallium is under investigation. Many reports of success and failure of gallium to demonstrate neoplastic lesions have appeared in recent literature (3–6).

Table 1. Radioactive Materials Used for Tumor Localization and the Mechanism Whereby They Concentrate in or Around Tumors

Mechanism	Material
Space-occupying lesion	Colloids
Physiological dysfunction	Halogens
Osseous reaction to injury	Polyphosphate
Precursor requirement	Amino acids
Blood brain barrior disruption	Sodium pertechnetate
Abnormal blood supply	Albumin
Tumor-specific antigen	Antibody
Unknown	Gallium, indium

Method of Study

Since 1968 we have investigated over 300 patients with a variety of neoplastic and nonneoplastic diseases. In recent months we have examined patients with ^{67}Ga and ^{111}In simultaneously. During this period of time we have not observed a single adverse reaction related to the administration of these radiopharmaceuticals. The initial studies were performed with gallium citrate which was prepared by heating gallium trichloride to dryness and resuspending in citrate solution. The solution was buffered to approximately pH7, and was injected intravenously as a rapid bolus after millipore filtration. Distribution of the radiopharmaceutical injected as a bolus was identical to that of a slow intravenous administration. We subsequently examined a gallium chloride intravenous injection and found that the biological distribution and efficacy as a tumor-localizing agent was identical to that of citrate. Subsequent to the intravenous injection there is a rapid association of radiopharmaceutical with serum proteins (7). A half-time clearance from the plasma of about 10 hours was observed. The clearance was in general more rapid in patients with large tumor burdens. Serum transferrin is the primary carrier protein for gallium. Experimental animal studies suggest that saturation of the transferrin with iron modifies the distribution of ^{67}Ga. Stable scandium increases the tumor/nontumor ration in tumor-bearing rats. Unfortunately, scandium in low doses causes red cell hemoloysis in humans.

We routinely perform our scans 48 hours after intravenous injection, which allows adequate time for blood background to drop and tumor concentrate to reach maximum. The patients are prepared for the study with a laxative on the night preceding the examination and an enema on the day of the study. Even with such preparation we often find excessive concentration of gallium in the abdominal region, which generally follows the contour of the colon. This region, which is clinically of great importance (especially in patients with lymphoma), may be so contaminated with radioactivity that the abdominal area is impossible to interpret. This is one of the major disadvantages of gallium as an abdominal tumor-localizing agent. Restudy of the patient at 72 hours occasionally reflects a shift in an area of radioactivity which suggests that the concentration was in fecal material as opposed to an abdominal mass, but we frequently see the same abdominal pattern of activity in the bowel at 72 hours. A mucosal excretory pathway is probable. The liver is easily visualized in a normal patient, and at times a small amount of activity is noted in the left upper quadrant, which is felt to be concentrated in the spleen. In many cases this may represent the splenic flexure of the colon. Large amounts of radioactivity in the left upper quadrant are distinctly abnormal, and may reflect splenic involvement with neoplastic disease such as Hodgkins' disease. A variable amount of radioactivity is noted in the bony structures of the pelvis and in the sacrum. The testes and the penis are usually seen, and in female patients the mucosa of the vagina is at times well defined. On posterior scanning the lower ends of the scapula can be seen, and should not be confused with intrathoracic lesions. In approximately the same area on the anterior scan one frequently notices accumulation of radioactivity in breast tissue, and the lactating breast has an even greater avidity for gallium (8). These are not to be confused with intrathoracic lesions. In summary, we evaluated a total body gallium scan for symmetry of distribution in normal sites, as well as appearance of radioactivity in abnormal sites. It is of importance to familiarize oneself with the normal distribution before attempting interpretation of a pathological scan.

A practical problem exists when one examines a patient with ^{67}Ga, which is related to the four photons that are present (Figure 1). The highest peak, 388 keV, constitutes only 8% of the total available photons, compared to the 93 keV peak which is 40% abundant. The two middle peaks, 184 and 296 keV, are 20 and 12%, respectively. We routinely examine patients with the total body rectilinear scanner peaked at 300 keV. Images are obtained with zero enhancement and zero background subtract. The anterior and posterior image is examined, and selected areas of interest are subsequently examined with the scintillation camera at the 93 keV peak. Patients receive 45 μCi of ^{67}Ga per kilogram of body weight.

CLINICAL APPLICATION—TUMOR-LOCALIZING PHARMACEUTICALS

Prior to the availability of ^{67}Ga we had investigated selenomethionine as a tumor-localizing agent in patients who were undergoing routine pancreatic scans. We therefore had the occasion to study a fairly large number of normal patients. We extended our studies to the examination of patients with a lymphoproliferative disease (Figure 2). Selenomethionine was considered less than ideal as a radiopharmaceutical because of its marked limitation relative to the dosimetry of selenium-75. We compared the results of ^{67}Ga scanning in patients we had previously studied with selenomethionine, and it become immediately obvious that ^{67}Ga was a better tumor-localizing agent.

As an aside, it may be important to establish that the basis upon which selenomethionine concentrates in tumors is well founded, and is related to protein synthesis rate. The mechanism of gallium is obscure. As a tumor-localizing agent, ^{75}Se selenomethionine or ^{75}Se selenocystine may prove to be as good as, if not a better than, either gallium or indium. One of the first patients we studied had massive involvement of the right shoulder and thorax with reticulum cell sarcoma (histiocytic lymphoma) (Figure 3). The patient had failed chemotherapy as well as radiation therapy when he was studied with ^{67}Ga. The accumulation of radioactivity in this patient's tumor mass exceeded the concentration of radioactivity in the liver and spleen. Biological distribution of gallium was clearly influenced by the tumor mass which had a striking avidity for the radiopharmaceutical. We were considering the administration of ^{66}Ga as a therapeutic modality in this patient, but unfortunately he succumbed to the disease prior to the availability of this radionuclide.

Hodgkin's and non-Hodgkin's lymphoma are of great interest to oncologists and radiation therapists, because they can cure Hodgkin's lymphoma by appropriately staging and treating involved areas with megavoltage therapy. Good results are predicted in early-stage Hodgkin's lymphoma but this is not the case with non-Hodgkin's lymphoma which seems to have been modified very little with the advent of either supervoltage theraphy or combined drug chemotherapy programs. Invasive and noninvasive techniques for staging have been investigated, and with our initial results in non-Hodgkin's lymphoma we actively sought patients in both disease categories. (Figures 4–9) Individual lesions were clearly defined in many cases, and approximately 80% of the patients with active lymphoma of both varieties had at least one positive site on scanning (Table 2). Some sites of proven disease, however, were not visualized. The retroperitoneal area, which is of great interest, was frequently impossible to evaluate because of contaminating activity in the colon. The reliability of lymphangiography as a diagnostic tool for retroperitioneal lymphoma has been recently questioned (9). Exploratory laparotomy appears to be the only acceptable modality for accurate staging.

A variety of neoplastic diseases other than lymphoma has been examined with varying degrees of success (Table 3). Both primary and metastatic lesions to the lung have almost uniformly been imaged successfully with ^{67}Ga (Figures 10 and 11). The size and site of the lesion influenced its detectability, as well as the biological

Table 2 Summary of Gallium Scan in Patients with Lymphoproliferative Diseases[a]

Histological Diagnosis	Number of Cases	Positive Scans
Malignant lymphoma (histiocytic)	15	11
Malignant lymphoma (lymphocytic)	9	6
Hodgkin's (mixed cellularity)	3	3
Hodgkin's (nodular sclerosis)	5	4
Hodgkin's (lymphocytic predominance)	2	1
Hodgkin's (lymphocytic depletion)	1	1
Hodgkin's (not classified)	17	12
Mycosis fungoides	3	1
Leukemic lymphosarcoma	1	1
Thymoma	1	1
Total	57	41

[a] Out of 57 patients 47 had at least one positive site on total-body scan.

Table 3 Summary of 172 Cases Studied Demonstrating Concentration of ^{67}Ga in Neoplastic and Nonneoplastic Lesions and Similar Results from Statistical Analysis of 300 Cases

Disease Process	Number of Patients	Positive Scans (%)
Neoplastic disease proven or probably active	129	70
Nonneoplastic diseases	21	38
Undefined illness (maybe neoplasm)	22	45
Malignant lymphoma	57	66
Untreated lymphoma	14	71
Proven or probably active disease	50	78

nature of the tumor per se (Figure 12). Tissues recovered at postmortem examination reflect a wide range of radioactivity per gram of tissue in the same patient (Table 4). This difference was not based on any obvious morphological difference.

Enthusiasm for ^{67}Ga total-body scanning as a diagnostic modality in a clinical oncology practice was somewhat dampened by the finding that nonneoplastic diseases modified rather dramatically the distribution of this material. Several patients with secondary hyperparathyroidism on renal dialysis demonstrated almost total concentration of the radionuclide into bone with almost no uptake in liver tissue (Fig. 13). Osteomyelitis concentrates gallium, as do pulmonary tuberculosis and sarcoidosis (Figures 14 and 15). Concern relative to concentration of gallium in nonneoplastic lesions, as well as difficulty in evaluating retroperitoneal disease because of excessive gut concentration, led us to seek other tumor-localizing radiopharmaceuticals. ^{111}In, which is in the same group in the periodic table as ^{67}Ga, was a logical material for investigation. Indium has more favorable radiation characteristics than gallium, and the two gamma peaks with excellent abundance give more than adequate numbers of suitable photons (10). Experimental animal studies have been performed comparing indium and gallium in both the chloride and citrate

Table 4 Variability Concentration of ^{67}Ga in Four Different Histological Types of Tumor[a]

Tumor	Counts per Minute per Gram			Tumor/ Muscle
	Minimum	Average	Maximum	
Poorly differentiated carcinoma, hepatocellular (necrotic)	3,429	5,536	8,267	0.76
Adenocarcinoma, lung	15,596	30,904	41,304	3.30
Histocytic lymphoma (reticulum cell sarcoma)	2,575	16,812	48,562	159.00
Squamous cell carcinoma, lung	8,552	18,151	33,828	52.00

[a] Grossly identical tumor tissue demonstrated wide ranges in concentration of radioactivity.

forms. Of the four chemical species, indium chloride has the greatest total body retention, as well as the highest tumor/muscle ratio in the particular tumor-bearing experimental animal we evaluated (11). In all animal experimentation one must exercise great caution in translating the information from the experimental animal to the human clinical situation. We have examined over 200 patients with indium chloride, and many patients have been examined sequentially with first indium and then gallium. In recent months we have examined patients with the simultaneous administration of gallium and indium. One major advantage of indium chloride is that there appears to be minimal concentration in the gut. Patients are not routinely prepared with laxatives or enemas, and the retroperitoneal area is usually quite clear. Bony uptake, which is striking, is predominantly marrow, and we have noted, as have other investigators, that irradiation or drugs, and at times replacement of marrow by tumor, is associated with an absence of indium concentration (Figures 16 and 17). The reticuloendothelial system in the area so affected may still be demonstrated to be intact by colloid scanning.

The clarity of ^{111}In scans become immediately obvious compared to ^{67}Ga in the same patient. The skeletal structure was seen quite clearly, and some tumors concentrated indium quite well (Figures 18 and 19). We evaluated several patients before and after therapy, and noted that the ^{111}In concentration disappeared from tumor sites that were adequately treated but appeared in the posttreatment scan at sites not adequately treated (Figure 20). Lymphangiographic determination in patients with a disease such as melanoma correlated quite nicely with the indium-111 study (Figure 21). Patients examined sequentially with ^{111}In and ^{67}Ga caused us some concern owing to the fact that lesions were seen at times less distinctly with indium than with gallium and at times not at all with indium (Figs. 22–24). The converse was never observed. Stated more specifically, we have never seen an ^{111}In-positive study that was ^{67}Ga-negative. We have seen ^{67}Ga-positive scans that were ^{111}In-negative. Both ^{111}In and ^{67}Ga fail to concentrate adequately in some lesions (Fig. 25). This failure is common with liver metastasis. Neither ^{111}In nor ^{67}Ga concentrates as well in an area defined as "cold" on technetium sulfur colloid as it does in normal surrounding liver tissue (Fig. 26).

Indium suffers from the same degree of nonspecificity as ^{67}Ga and concentrates quite avidly in osteomyelitis (Figure 27). It is of great interest to us that scanning with ^{111}In, ^{67}Ga, and ^{99m}Tc polyphosphate in patients with metastatic tumor to bone, the ^{67}Ga and ^{99m}Tc polyphosphate studies correlate, whereas the indium-111 study reflects a disruption of the normal concentration of activity in the bone marrow. This is especially common in the posterior pelvic region. We examined this butterfly area of activity and have found a surprising large number of patients who have an abnormal pattern in this bony area that corresponds to neoplastic disease involvement (Figures 28–30).

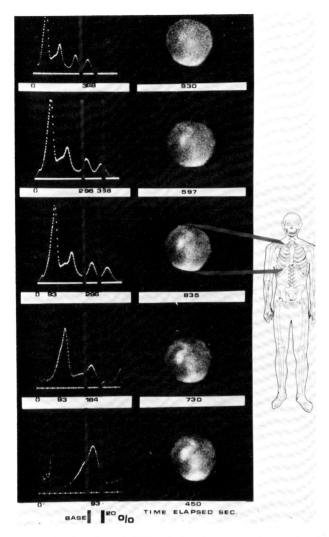

Fig. 1. Camera images of an intrathoracic histiocytic lymphoma presented as a function of the time required for the image, as well as the photo peaks examined. The 93 keV peak resulted in an acceptable image in the least time.

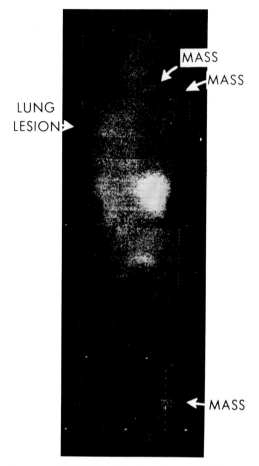

Fig. 2. Total-body posterior scan with [^{75}Se]methionine reflects multiple cutaneous as well as lung lesions. Reticulum cell sarcoma with multiple subcutaneous mass.

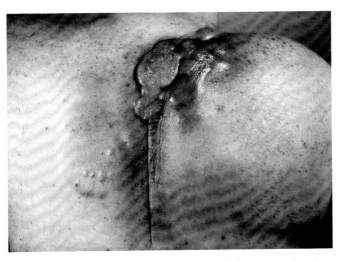

Fig. 3A. Photograph of massive involvement of right shoulder with histiocytic lymphoma.

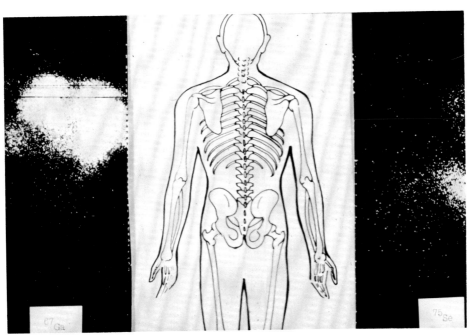

Fig. 3B. [^{75}Se]methionine and ^{67}Ga in same patient, different times. Reticulum cell sarcoma.

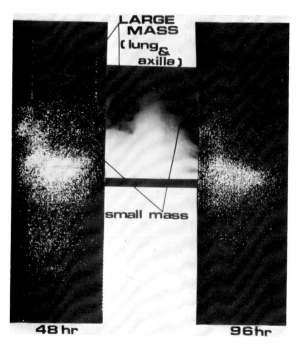

Fig. 4. Rectilinear scan at 48 and 96 hours demonstrates lesion in right lung behind pleural effusion, and smaller lesion in left lung which is also seen on x ray. Scintillation camera images at 48 hours seen in Figure 1. ^{67}Ga total-body scan. Reticulum cell sarcoma.

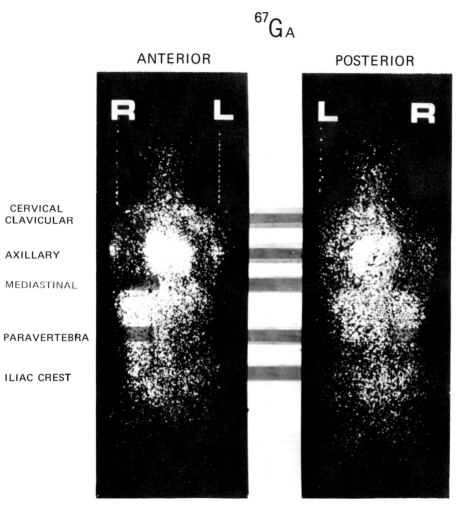

Fig. 5. Anterior and posterior rectilinear scans demonstrating multiple sites of involvement with massive mediastinal lesion. Reticulum cell sarcoma.

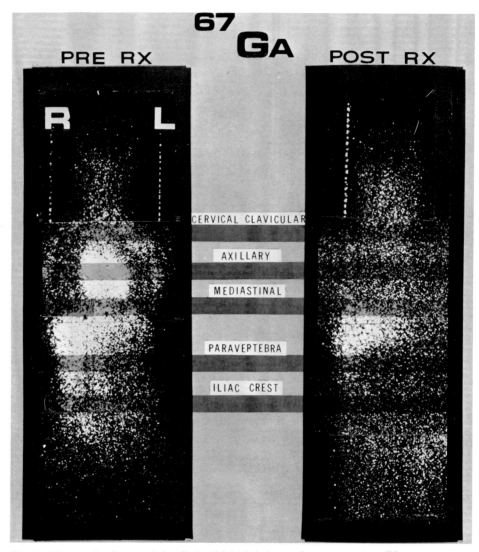

Fig. 6. Three weeks after mantle irradiation. Multiple lesions no longer concentrate ^{67}Ga.

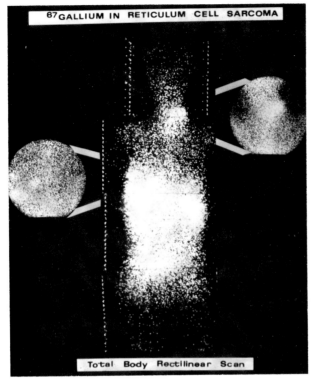

Fig. 7. Rectilinear and camera images of an achondroplastic dwarf with histiolytic lymphoma. Neck mass and lung lesions clearly defined.

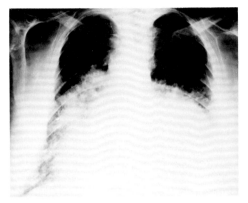

Fig. 8A. X-ray of a patient with known Hodgkin's mixed cellularity.

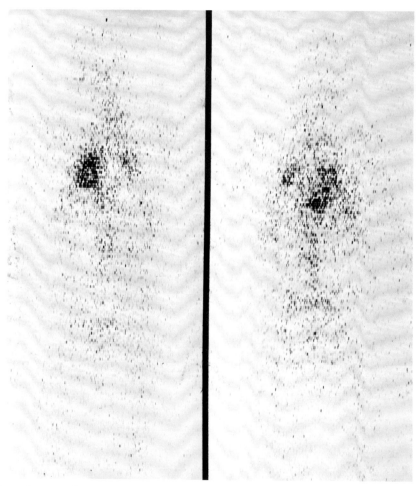

Fig. 8B. Corresponding total-body rectilinear anterior and posterior ^{67}Ga scan. E.S. 1045925.

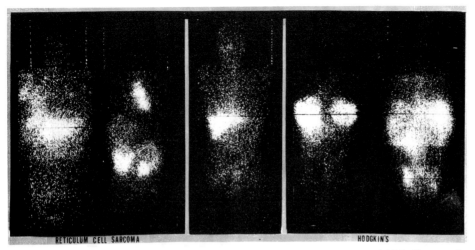

Fig. 9. Summary slide of ^{67}Ga in lymphoma; central figure is 7 year postcure Hodgkin's disease to be compared with two active cases of Hodgkin's disease on the right and two cases of histiocytic lymphoma on the left. Patients with Hodgkin's disease were found to have splenic involvement at time of exploration.

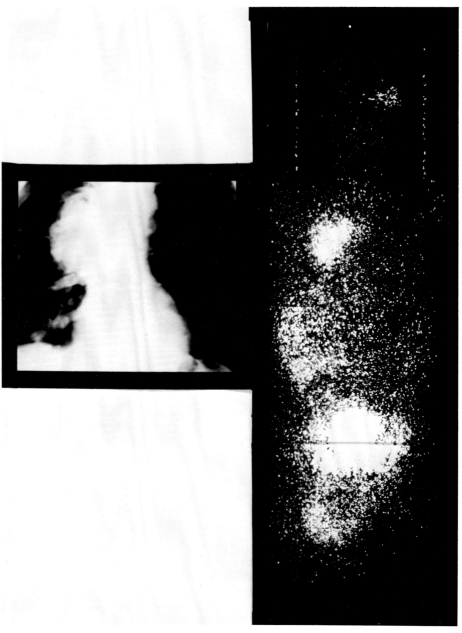

Fig. 10. Chest x ray reveals right hilar mass, and total-body scan defines the chest lesion. In addition, exophytic skull, abdominal, and pelvic lesions are noted. Neuroblastoma.

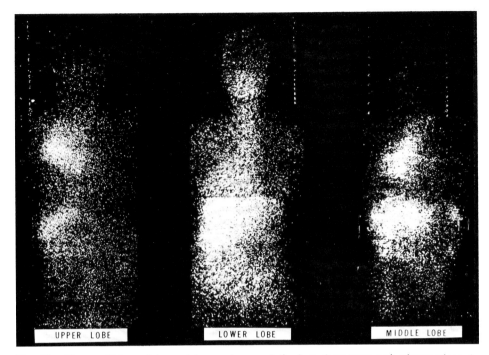

Fig. 11. Upper middle and lower lobe carcinoma of the lung (squamous and adenocarcinoma) concentrate ^{67}Ga.

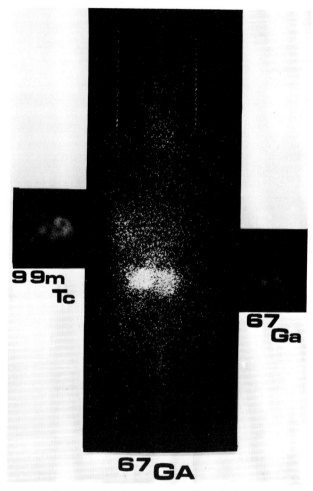

Fig. 12. Leiomyosarcoma is seen in the abdomen with gallium-67, but the area of involvement in the liver as defined with technetium colloid as filling defect remains cold.

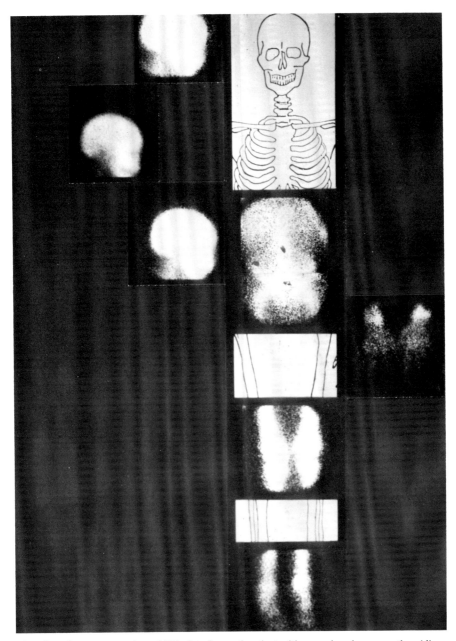

Fig. 13. Complete concentration of ^{67}Ga into bone of patient with secondary hyperparathyroidism on renal dialysis. M.S. N-9202.

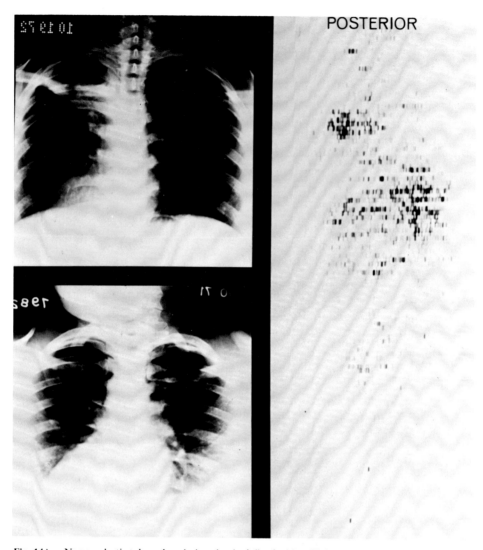

Fig. 14A. Nonneoplastic tuberculous lesion clearly defined with gallium-67. L.M. 798235.

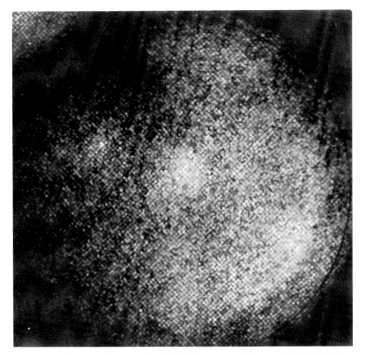

Fig. 14B. Same patient seen in scintillation camera image. L.M. 798235.

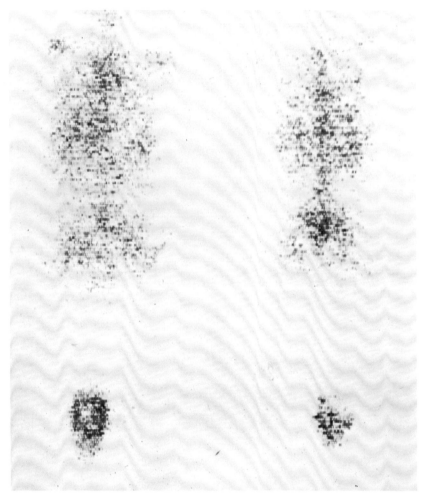

Fig. 15. Osteomyelitis of right femur concentrates gallium-67. E.F. 685952.

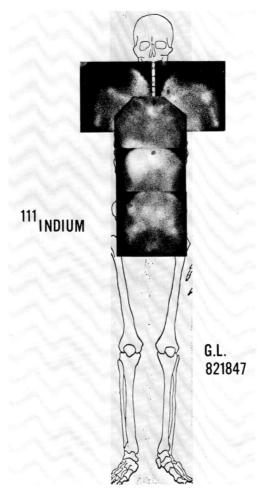

Fig. 16. Indium-111 marrow scan in patient with Hodgkin's disease and previous radiation therapy to right axilla, which suppressed marrow in right humerus.

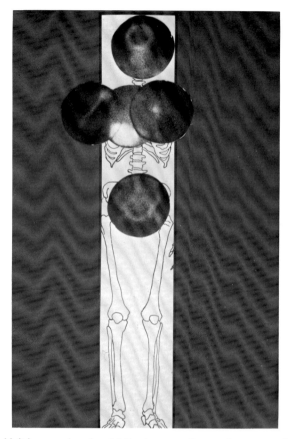

Fig. 17. Marrow of left humerus is reduced following x-ray therapy, but metastatic lymph node in left axilla is clearly defined with ^{111}In. A.P. 864485.

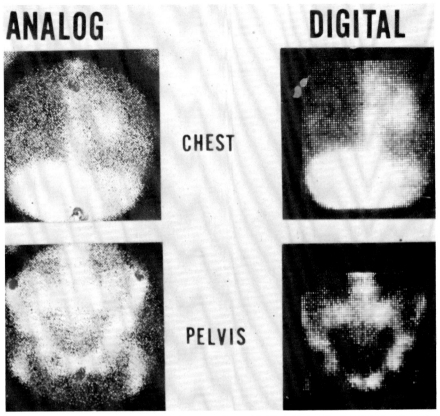

Fig. 18. Digital and analog images of ^{111}In in patient with Hodgkin's mixed cellularity. Lung lesion vividly demonstrated. N.T. N-9671.

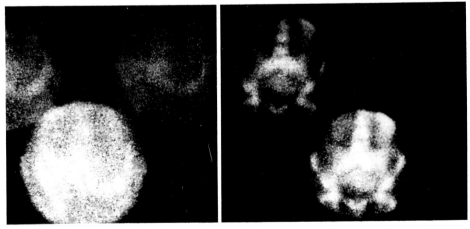

Fig. 19. Comparison of ^{67}Ga (left) and ^{111}In (right) in same patient. The pelvic structures (anterior view) are distinctly visualized on the indium scan.

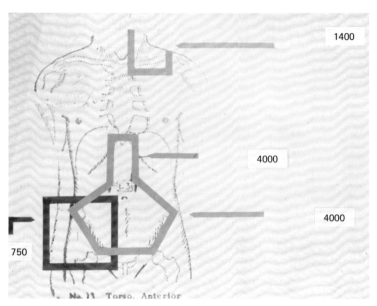

Fig. 20A. Radiation fields in patient with Hodgkin's mixed cellularity. Pancytopenia caused cessation of therapy at this time. Right inguin area received 4000 plus 750 R. Left supraclavicular region only 1400 R.

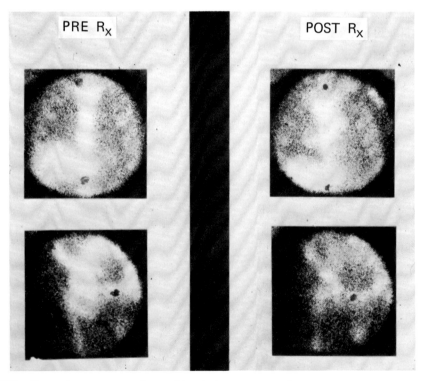

Fig. 20B. Pre- and posttreatment chest and inguinal area scans. Indium appears to concentrate in new lung lesions and left supraclavicular area, but abnormal accumulation in right femoral area disappears. C.B. 1004328.

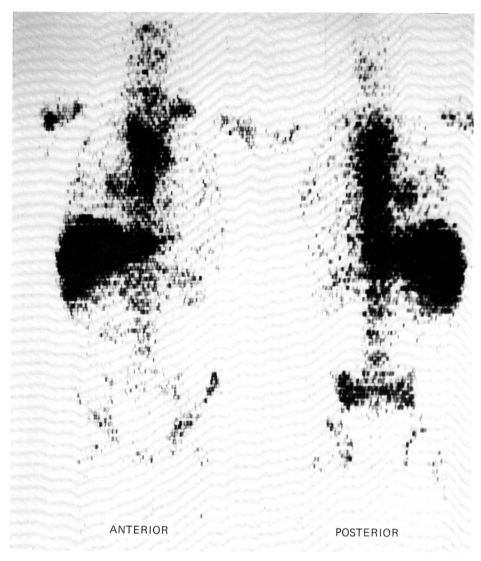

Fig. 20C. Rectilinear scan, anterior and posterior, of same patient demonstrating left supraclavicular, mediastinal, and hilar nodes. C.B. 1004328.

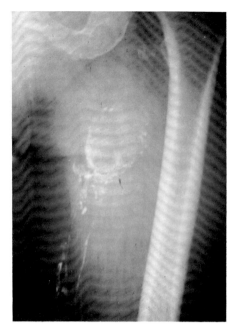

Fig. 21A. Lymphangiogram in patient with melanoma.

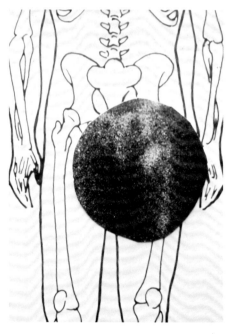

Fig. 21B. Corresponding indium scan.

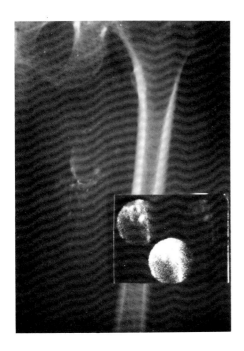

Fig. 21C. Composite of two studies demonstrating positive node on lymphangiogram and positive indium scan.

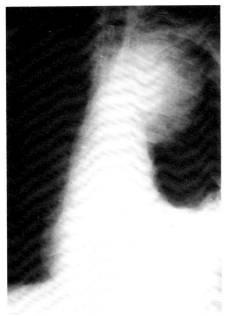

Fig. 22A. Large left upper lobe lesion seen on routine chest x ray.

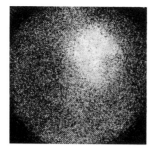
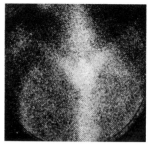

Fig. 22B. Scintillation camera images with gallium-67 (left) and indium-111 (right) demonstrate striking uptake of gallium and negligible concentration of indium-111.

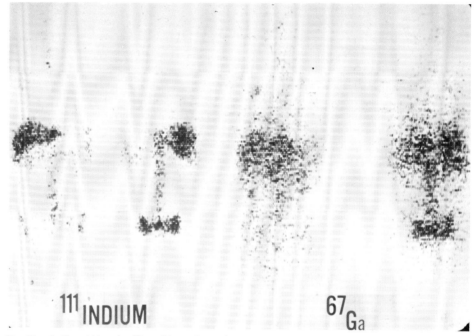

Fig. 23. Indium-111 and gallium-67 anterior and posterior rectilinear scanning lesion in right lung seen faintly with indium, quite well with gallium.

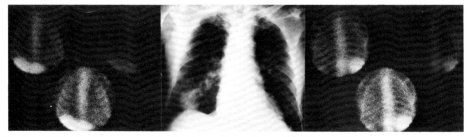

Fig. 24. Right lung lesion, not defined with anterior and posterior ^{111}In scintillation camera images.

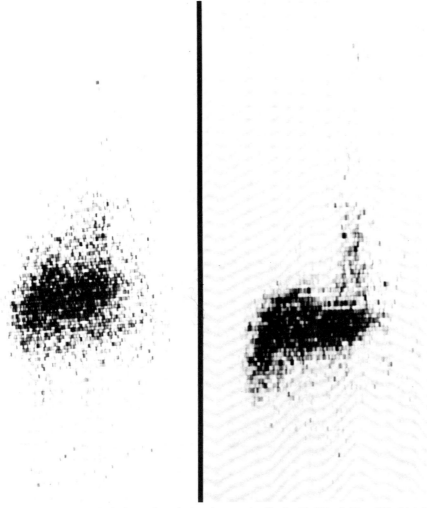

Fig. 25. Massive left supraclavicular and cervical nodes not visualized with ^{67}Ga (left) or ^{111}In (right). Anterior scan. Proven Hodgkin's disease. J.C. N-10082.

 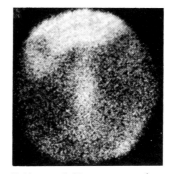

Fig. 26. Liver metastasis demonstrated on technetium colloid scan (left) not seen as hot areas on indium-111 scan (right). Metastatic cancer to liver. F.P. N-9704.

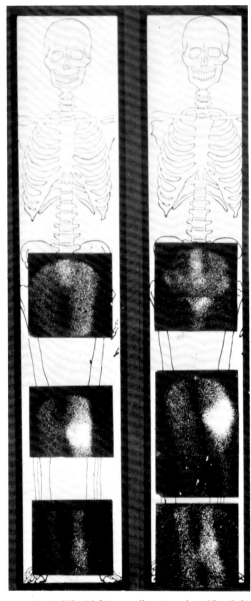

Fig. 27. Osteomyelitis concentrates ^{111}In (right) as well as strontium-87m (left). J.C. 990022.

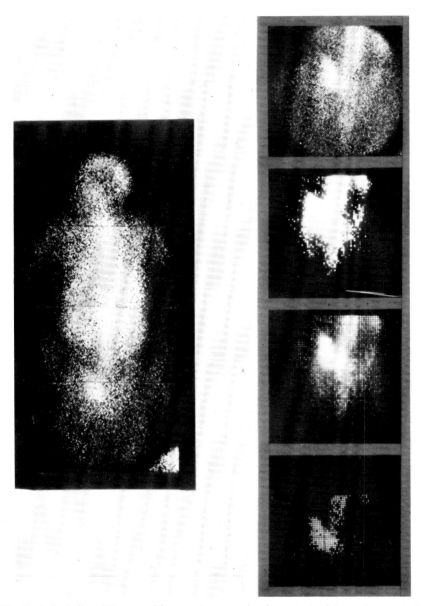

Fig. 28. Posterior indium-111 scans with computer processing demonstrate absence of concentration of ^{111}In in right sacroiliac region. Patient proven to have tumor involvement in this region. R.F. N-9757.

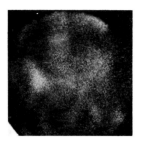  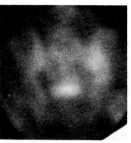

Fig. 29. Comparison of indium-111 (left), gallium-67 (center), and technetium polyphosphate (right) images. Technetium polyphosphate and gallium reflect increased concentration in area of tumor involvement. Indium-111 shows decreased marrow activity.

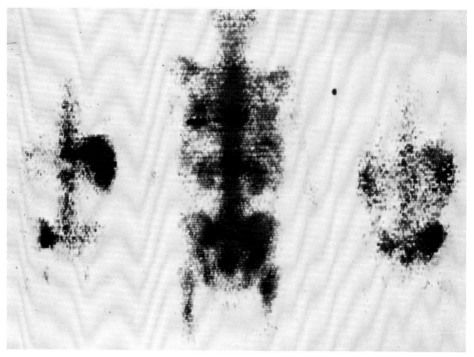

Fig. 30. ^{111}In (left), ^{99m}Tc polyphosphate (center), and ^{67}Ga (right) posterior scans of the same patient. Rectilinear images reveal decreased ^{111}In concentration in area of increased technetium polyphosphate and increased ^{67}Ga.

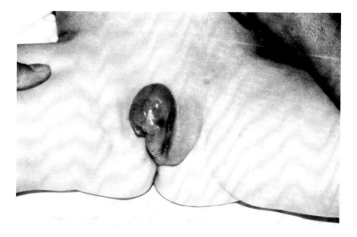

Fig. 31A. Photograph of Burkitt's lymphoma of right labia.

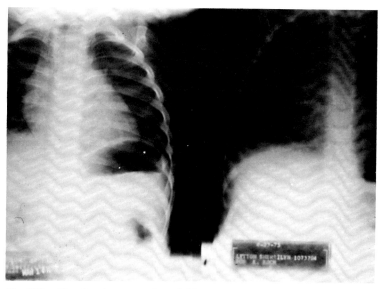

Fig. 31B. Negative chest x-ray time of ^{67}Ga scan.

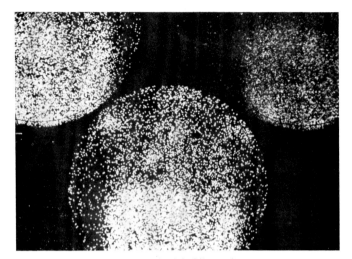

Fig. 31C. Positive gallium-67 in right hilar region.

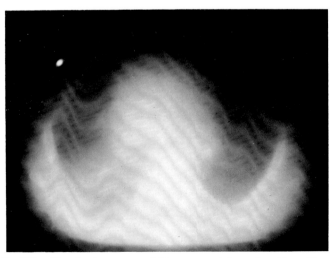

Fig. 31D. Negative transverse axial-tomogram at time of positive gallium scan.

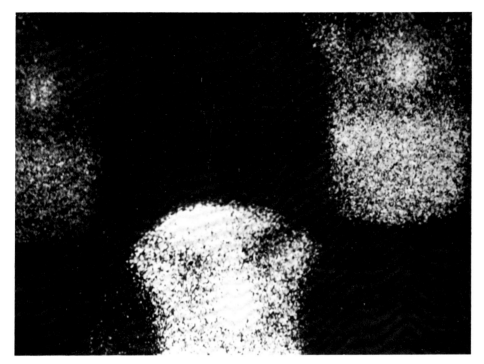

Fig. 31E. Scintillation camera image 3 months later with marked increase in size of hilar lesion.

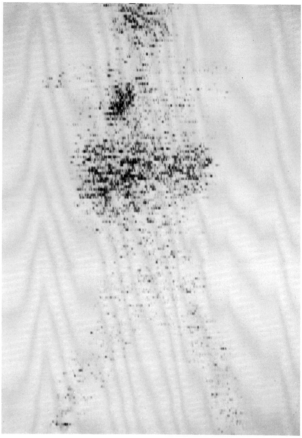

Fig. 31F. Total-body rectilinear scan at same time of second camera study, demonstrating right hilar lesion.

SUMMARY

Total-body scanning for tumor localization is a procedure that is receiving increasing attention, primarily, because no mobidity or mortality is associated with it. Requests to the nuclear medicine laboratory should be phrased in the context of questions such as: Does my patient have an easily accessible mass for biopsy? or, does his disease exist below the diaphragm? In posing any questions, one must be aware of the fact that tumor-localizing radiopharmaceuticals concentrate in non-neoplastic disease, and that a patient with carcinoma may well have tuberculosis or osteomyelitis as a second disease.

We recently evaluated an infant with Burkitt's lymphoma of the right labia. This region was irradiated and subsequently the patient was evaluated by total-body scan. We noted an area of positive uptake at the right hilum, which a chest film failed to reflect. The patient received no further therapy, since it was felt to be a false positive scan. The patient was rescanned 3 months later, and it was found that the chest lesion had increased in size (Figure 31). The patient was at this time placed on Cytoxan and the lesion disappeared. Positive gallium scans in areas where other modalities fail to reflect disease can be interpreted as (1) false positive, or (2) detection of a pathophysiological process which is unrecognizable by other diagnostic means. We anticipate continued follow-up of our patients who have positive scans without other evidence of disease, because we feel that this may be the best measure of the value of a total-body gallium scan.

REFERENCES

1. Edwards, C. L. and Hayes, R. L.: Tumor scanning with ^{67}Ga citrate. *J. Nucl. Med.* **10**:103–105, 1969.
2. Swartzendruber, D. C., Nelson, B., and Hayes, R. L.: Gallium-67 localization in lysosomal like granules of leukemic and non leukemic murine tissues, *J. Nat. Cancer Inst.* **46**:942–952, 1971.
3. Lavender, J. P., Lowe, J., Barker, J. R., Burn, J. I., and Chaudhri, M. A.: Gallium-67 citrate scanning in neoplastic and inflammatory lesions. *Brit. J. Radiol.* **44**:361–366, 1971.
4. Vaidya, S. G., Chaudhri, M. A., Morrison, R., and Whait, D.: Localization of gallium-67 in malignant neoplasms. *Lancet* **2**:911–941, 31, 1970.
5. Langhammer, H., Glaubitt, G., Grebe, S. F., Hampe, J. F., Haubold, U., Hor, G., Kaul, A., Koeppe, P., Koppenhagen, J., Roedler, H. D., and Van der Schoot, J. B.: ^{67}Ga for tumor scanning. *J. Nucl. Med.* **13**:15–30, 1972.
6. Turner, D. A., Pinsky, S. M., Gottschalk, A., Hoffer, P. B., Ultmann, J. E., and Harper, P. U.: The use of ^{67}Ga scanning in the staging of Hodgkin's disease. *Radiology* **104**:97–101, 1972.
7. Hartman, R. E. and Hanes, R. L., Gallium binding by blood serum. *Fed. Proc.* **26**:780, 1967; *J. Pharmacol. Exp. Ther.* **168**:193, 1969.
8. Fogh, J. and Thorn, N. A.: ^{67}Ga accumulation in malignant tumors and in the prelactating or lactating breast. *Proc. Soc. Exp. Biol. Med.* **138**:1086–1090, 1971.
9. Glatstein, E., Trueblood, H. W., Enright, L. P., Rosenberg, S. A., and Kaplan, H. S.: Surgical staging of abdominal involvement in unselected patients with Hodgkin's disease. *Radiology* **97**:425, 1970.
10. Goodwin, D. A., Goode, R., Brown, L., and Imbornone, C. J.: ^{111}In labeled transferring for the detection of tumors. *Radiology* **100**:175–179, 1971.
11. Weinstein, M. B.: ^{67}Galliuum and ^{111}indium concentration in R2788 lymphosarcoma (rodent). Unpublished data.

Labeled Chloroquine Analog in Diagnosis of Ocular and Dermal Melanomas

CHAPTER FOURTEEN   WILLIAM H. BEIERWALTES, M.D.

DEVELOPMENTAL BACKGROUND

The idea for our development of a radioiodinated analog of chloroquine originated from the observation that chloroquine used to treat malaria occasionally caused a retinopathy. Several reports indicated that chloroquine has a marked affinity for melanin and is slowly released from pigmented tissues (1–3).

The results of our feasibility study are shown in Table 1 (4). The relative tissue concentrations of ^{14}C chloroquine were studied in 32 mice with melanomas, 16 black and 16 albino, at 1, 2, 3, and 4 days after injection. Radioactivity increased through 4 days in the eyes of black mice (containing melanin in the choroid), but decreased rapidly in the eyes of albino mice (without melanin in their choroids). Of all other tissues studied, the only tissues showing a relatively constant high activity throughout the 4 days were melanin-containing malignant melanotic melanomas in both pigmented and albino mice and in the skin of black (but not albino) mice.

Figure 1 shows the structure of the ^{14}C-labeled chloroquine and the structure of the iodochloroquine analog 4-(3-dimethylaminopropylamino) ^{125}I-7-iodoquinoline that our colleague, Raymond Counsell, synthesized (5,6).

The tissue distribution of radioactivity was then compared with that of ^{14}C chloroquine under similar conditions in 12 mice and in 5 hamsters with melanomas (4). The distribution of the same quantities of ^{125}I from $Na^{125}I$ was studied as a control in 12 pigmented mice with melanoma, and in 2 hamsters with melanoma. Further control animals consisted of 2 hamsters with melanomas given 100 vials of ^{131}I as radioiodinated serum albumin (RISA) and, in 2 hamsters, 100 μCi of ^{197}Hg as chlormerodrin. The data demonstrated that a derivative of chloroquine had been synthesized and labeled with radioactive iodine without destroying the specificity of this quinoline for melanin. The concentration ratio of ^{125}I from this radioiodinated analog (NM-113) was sufficient in every instance to delineate the malignant melanoma by scintillation scanning.

We then studied the diagnostic efficacy of this compound in three patients. Figure 2 is a picture of the right side of the face of an 83 year old woman. Note the

Table 1 ^{14}C Chloroquine Tissue Distribution in Mice with Melanomas[a]

Time after injection (hours)	Melanoma		Eye		Skin		Liver	
	B-16	H.P.	B-16	H.P.	B-16	H.P.	B-16	H.P.
25	424 ± 17	583 ± 410	911 ± 83	106 ± 26	237 ± 49	103 ± 49	222 ± 24	237 ± 34
48	445 ± 58	279 ± 73	954 ± 132	35 ± 2.4	265 ± 40	46 ± 4.6	100 ± 15	138 ± 7
72	376 ± 38.3	178 ± 36	1086 ± 70	13 ± 6.3	260 ± 4	57 ± 12	41 ± 7.1	29 ± 1.7
96	379 ± 31.6	359 ± 258	1479 ± 239	44 ± 20	352 ± 32.6	28 ± 3.9	30 ± 3.5	31 ± 5.6

[a] Four mice were used in each experiment; values are given in counts per minute per milligram. B-16 melanomas in pigmented mice; H.P. melanomas in albino mice.

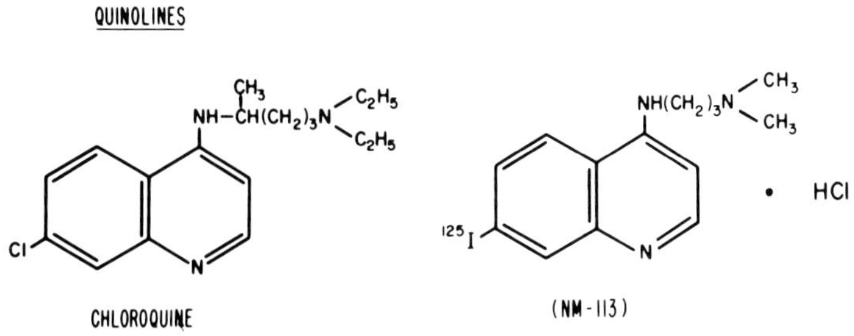

DEVELOPMENTAL BACKGROUND

QUINOLINES

Fig. 1. Structure of ^{14}C-labeled chloroquine and the iodochloroquine analog 4-(3-dimethylaminopropylamino ^{125}I-7-iodoquinoline.

primary lesion in the right cheek with satellites posterior to it, a nonvisible but palpable preauricular node; a visible black satellite below and behind the ear; and a neck dissection scar without visible or palpable lesions.

Figure 3 is a photograph of the normal left side of her face. There was a palpable "lymph node" at the angle of the jaw similar to that found on the right side of the face.

Figure 4a is a photograph of the scintillation scan of the right side of her face. Note the considerable concentration of radioactivity in the region of the preauricular node (which proved to contain metastatic melanoma showing 565 counts/minute of tissue). Also evident is concentration in the obvious black satellite behind the ear and in the region of melanoma containing lymph nodes deep to the neck dissection scar.

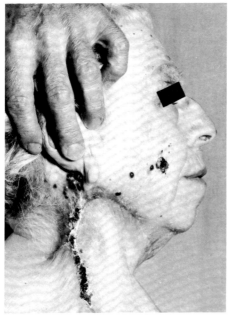

Fig. 2. Picture of the right side of the face of an 83 year old woman. Note the primary lesion in the right cheek with satellites posterior to it; nonvisible but palpable preauricular node; visible black satellite below and behind the ear and neck dissection scar without visible or palpable lesions.

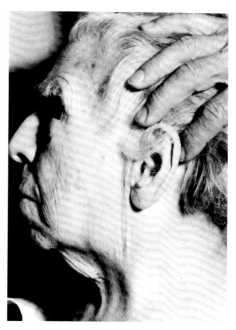

Fig. 3. Photograph of the normal left side of her face. There was a palpable "lymph node" at the angle of the jaw similar to that found on the right side of the face.

Figure 4b is a photograph of the scintillation scan of the left side of her face. No significant concentration of radioactivity is seen in the palpable preauricular "lymph node," which proved to be "chronic inflammation with parotid gland, no melanoma," with one-tenth the radioactivity in counts per minute per milligram of tissue.

Our next and last publication on the diagnostic efficacy of NM-113 in dermal melanomas summarized our experience in 30 humans (8). The radioactivity concentration ratio for melanoma compared to skin, muscle, and fat ranged from approximately 7:1 to 60:1. Amelanotic tissue and areas of necrosis in melanotic tissue concentrated the compound poorly, if at all, compared with other tissue. No false positive scans were obtained, but there were false negative scans in five patients. In one patient the false negative scans were due to the small size of the lesion. The remaining four patients had amelanotic lesions.

The limitations and indications for the use of NM-113 might be summarized as follows:

Limitations

1. Uptake in brain metastases is nonspecific.
2. Uptake in normal lungs is too great to allow detection of metastasis to lung.
3. Uptake in lungs prevents accurate imaging of axillary nodes.
4. Radioactivity excretion in bile and bowel prevents adequate imaging of retroperitioneal metastasis.
5. Cannot detect lesions smaller than 1 cm in inguinal nodal areas.
6. No uptake in amelanotic melanoma.

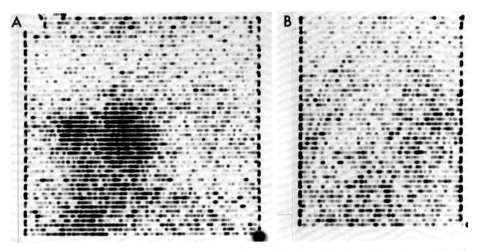

Fig. 4. (A) Photograph of the scintillation scan of the right side of her face. Note the considerable concentration of radioactivity in the region of the preauricular node (which proved to contain metastatic melanoma showing 565 counts/minute per milligram of tissue.) Also evident is concentration in the obvious black satellite behind the ear, and in the region of the melanoma containing lymph nodes deep to the neck dissection scar. (B) Photograph of the scintillation scan of the left side of her face. No significant concentration of radioactivity is seen in the palpable preauricular "lymph node," which proved to be "chronic inflammation with parotid gland. No melanoma," with one-tenth the radioactivity concentration.

Indications:

1. Can detect presence of melanotic melanoma metastasis in regional lymph nodes in face and neck, even when the nodes are deep cervical and nonpalpable.
2. Can differentiate melanotic melanoma metastastic to liver from nonmelanotic melanoma (as shown in Figure 5).

DIAGNOSIS OF OCULAR MELANOMA

The compound has found its best use in the diagnosis of ocular melanoma. Here, Ferry has shown that 19% of enucleated eyes with a diagnosis of malignant melanoma do *not* contain a tumor (9). The ^{32}P test is not specific for ocular mel-

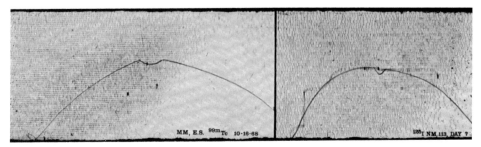

Fig. 5. Metastatic malignant melanoma of the left hepatic lobe. (Left) ^{99m}Tc liver scan showing cold defect in left lobe. (right) Left upper quadrant abdominal scans with [^{125}I]-NM-113 at day 10 after dose shows positive uptake of radioactivity from [^{125}I]-NM-113 in region of tumor where ^{99m}Tc scan shows a lack of radioactivity concentration. (From color scan).

anoma, and it has shown both false positive and false negative results. Although the direct placement of the probe over the tumor in the operating room is said to result in fewer enucleations for benign lesions (10), these tragic mistaken diagnoses still occur regularly.

Figure 6 is a photograph of a horizontal section of an eye removed because it contained a 15 × 11 mm diameter malignant melanotic melanoma. Melanomas this size are almost always symptomatic. Figure 7, however, is a photograph of a horizontal section of an eye containing an asymptomatic 7 mm diameter subretinal hemorrhage at the back of the globe, which gave a "diagnostically high" uptake count with ^{32}P in the operating room under direct probe placement.

Figure 8 is a graph showing the results of our first efforts (11) with NM-113 in imaging ocular melanomas in 28 patients. Radioactivity in each of the two eyes of each patient was compared by external counting using a specially designed 1 in. single-hole lead collimator on a 5 in. NaI (tl) crystal photoscanner with an attached scaler. Radioactivity determinations by external counting were made one to five times in each patient from 2 to 50 days after the tracer dose. Results were expressed as the average percent difference between the mean counting rates over two eyes using the formula:

$$\frac{A - B}{(A + B)/2} \times 100$$

Although relatively large melanomas were detected with this relatively crude geometry, it occurred to us that if we could decrease the mean percent difference in external counting rate between the eyes of 7.6% after 14 days following the dose with an upper limit of 18% with 2 s.d., we could detect much smaller melanomas.

Figure 9 is a graph showing the importance of detecting small ocular melanomas (12). The mortality rate is threefold greater at a tumor volume of 2.2 cc as compared to 1.2 cc.

Figure 10 is a picture of the eye probe that our colleague, Glen Knoll, designed and fabricated especially for this problem with ^{125}I (13).

Figure 11 shows that, by using several procedures to assure a reproducible

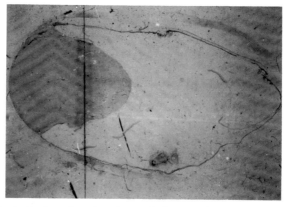

Fig. 6. Photograph of a horizontal section of an eyeball removed because it contained at 15 × 11 mm diameter malignant melanotic melanoma.

DIAGNOSIS OF OCULAR MELANOMA

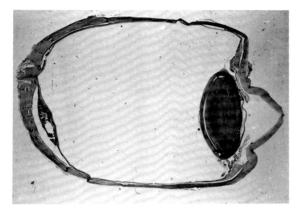

Fig. 7. Photograph of horizontal section of an eye containing a 7 mm diameter subretinal hemorrhage at the back of the eye which gave a "diagnostically high uptake" count with ^{32}P in the operating room under direct probe placement.

geometry with this probe, we were able to decrease the difference in counting rates between the two eyes to $< 6\%$ between days 5 and 22 after the tracer dose.

Now that we are in a position to detect much smaller differences in count rates between the two eyes attributable to smaller malignant melanotic ocular melanomas, the problem we confront is finding this population of patients before their tumors become large enough to produce pain and gross disturbances in vision.

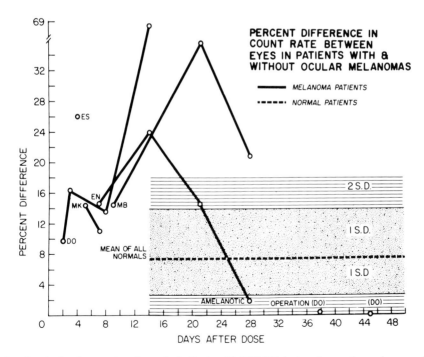

Fig. 8. Graph showing results of our first efforts with NM-113 in imaging ocular melanomas in 28 patients.

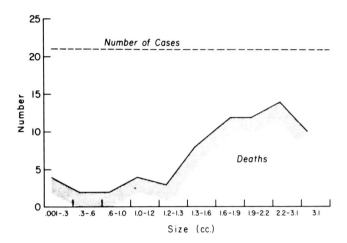

Fig. 9. Graph showing the relationship of mortality rate to ocular melanoma size.

There are two populations of ophthalmology patients who deserve close scrutiny because of their high risk of having an ocular melanoma: (1) those having routine refraction where the opthalmologist sees something that involves a differential diagnosis of choroidal nevus, retinal or choroidal hemangioma, retinal hemorrhage, retinal or choroidal detachment, or retinal or macular degeneration; (2) where opacities present adequate fundoscopy: corneal, lenticular, or vitreous opacities, secondary membranes, degenerated eye, and ptesis bulbi; these blind eyes with opaque media, for some unknown reason, harbor melanomas with disturbing frequency (14).

The diagnostic dose of 2 mCi usually used in this study would not be expected to produce radiation damage of the retina. Radiation dosimetry shows that the choroidal dose is approximately 46 rads. Harmful permanent effects on the retina from external radiation at much higher dose rates have been seen only with over 3250–4500 rads (15). Furthermore, in our studies of the treatment of dermal melanotic melanomas in dogs (16), we have shown that it would require a whole-body dose of three times the LD $_{50/30}$ to reach a damaging dose of beta radiation to the

Fig. 10. Picture of our new eye probe designed and fabricated by our colleague, Glen Knoll, especially for eye tumor localization with [^{125}I]-NM-113.

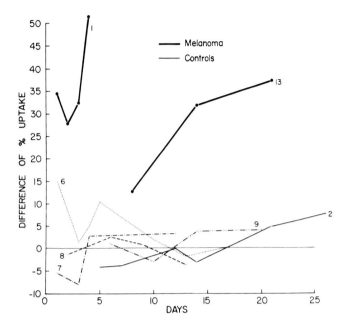

Fig. 11. Graph showing that this new probe achieved a normal difference in count rate between the two eyes of less than 6% between days 5 and 22.

rod and cone layer of the retina (17). In unpublished work, we have treated three humans with 25 mCi, three with 50 mCi, and one with 100 mCi of NM-113 and followed their retinal function for over 3 months with methods more sensitive than testing the visual acuity. No impairment of retinal rod or cone function followed the use of these therapy doses.

* Supported in part by USPHS Training Grant CA-05134-09 and -10, USAEC Grant AT (11-1) 2031, the Elsa U. Pardee Foundation, Nuclear Medicine Research Fund, and American Cancer Society Grant PRA-18.

REFERENCES

1. Bernstein, H., Zvaifler, N., Rubin, M., and Mansour, A. M.: The ocular deposition of chloroquine. *Invest. Ophthalmol.* **2**:384, 1963.
2. Sams, W. M., Jr., and Epstein, U. W.: The affinity of melanin for chloroquine. *J. Invest. Ophthalmol.* **45**:482, 1965.
3. Potts, A. M.: The reaction of uveal pigment *in vitro* with polycyclic compounds. *Invest. Ophthalmol.* **3**:405, 1964.
4. Beierwaltes, W. H., Varma, V. M., Lieberman, L. M., Counsell, R. E., and Morales, J. O.: Scintillation scanning of malignant melanomas with radioiodinated quinoline derivitives. *J. Lab. Clin. Med.* **72**:485, 1968.
5. Counsell, R. E., Poche, P., Morales, J. O., and Beierwaltes, W. H.: Tumor localizing agents. III. Radioiodinated quinoline derivitives. *J. Pharm. Sci.* 1042, 1967.

6. Counsell, R. E., Pocha, P., Ranade, V. V., Steingold, J., and Beierwaltes, W. H.: Tumor localizing agents. VII. Radioiodinated quinoline derivitives. *J. Med. Chem.* **12**:232, 1969.
7. Beierwaltes, W. H., Lieberman, L. M., Varma, V. M. and Counsell, R. E: Visualizing human malignant melanoma and metastases. Use of chloroquine analog tagged with iodine 125. *J. Amer. Med. Assoc.* **206**:97, 1968.
8. Boyd, C. M., Lieberman, L. M., Beierwaltes, W. H., and Varma, V. M.: Diagnostic efficacy of a radioiodinated chloroquine analog in patients with malignant melanoma. *J. Nucl. Med.* **11**:479, 1970.
9. Ferry, A. P.: Lesions mistaken for malignant melanoma of the posterior uvea, a clinicopathologic analysis of 100 cases with ophthalmoscopically visible lesions. *Arch. Ophthalmol.* (Chicago) **72**:463, 1964.
10. Hagler, W. S., Jarrett, W. H. and Humphrey, W. T.: The radioactive phosphorous uptake test in diagnosis of uveal melanoma. *Arch. Ophthalmol.* (Chicago) **83**:548, 1970.
11. Boyd, C. A., Beierwaltes, W. H., Lieberman, L. M., and Bergstran, T. J.: ^{125}I labeled chloroquine analog in the diagnosis of ocular melanomas. *J. Nucl. Med.* **12**:601, 1971.
12. Flocks, M., Gerende, J. H., and Zimmerman, L. E.: The size and shape of malignant melanomas of the choroid and ciliary body in relation to prognosis and histologic characteristics (a statistical study of 210 tumors). *Trans Amer. Acad. Ophthalmol. Otohinol.* **59**:740–758, 1955.
13. Knoll, G. F., Lieberman, L. M., Nishiyama, H., and Beierwaltes, W. H.: A gamma ray probe for the detection of ocular melanomas. *IEEE Trans. Nucl. Sci.* NS-**19**:76–80, February 1972.
14. Hogan, M. J. and Zimmerman, L. E. (eds.): *Ophthalmic Pathology,* 2nd ed. W. B. Saunders Company, Philadelphia, 1962, p. 429.
15. Reese, A. B., Merriam, G. R., and Martin, H. E.: Treatment of bilateral retinal blastoma by radiation and surgery: Report on 15 year results. *Amer. J. Ophthalmol.* **32**:175, 1949.
16. Lieberman, L. M., Boyd, C. M., Varma, V. M., Bergstram, T. J., and Beierwaltes, W. H.: Treatment doses of I-131 labeled chloroquine analog in normal and malignant melanoma dogs. *J. Nucl. Med.* **12**:153, 1971.
17. Lieberman, L. M.: The effects of radiation on the retina of the dog. Doctoral dissertation, University of Michigan, Ann Arbor, Mich., 1970.

Radiolabeled Amino Acids, Antigens, and Organic Compounds in Tumor Localization*

CHAPTER FIFTEEN   RICHARD P. SPENCER, M.D., Ph.D.

Most biological systems can be overly simplified as being made up of a mass of carbon compounds in a media of electrolytes. The central problem is then to distinguish one assemblage of the system, considered normal, from abnormal configurations. The obvious approach would be to use carbon-labeled compounds if—and this is a big if—a carbon radionuclide were available that had a half-life of several hours and which emitted a gamma ray of from 100 to 500 keV. Since no such carbon radionuclide exists, we must rely on indirect approaches. I first deal with four techniques presently available to us for following tumors. Attention will then turn to newer techniques under development.

TECHNIQUES

1. *Use of flow studies.* Radionuclide cerebral blood flow studies have become a useful first step prior to static brain scans. The suggestion has been made that a cerebral blood flow study should precede each brain scan (1). Indeed, we have been doing this for several years, and have found multiple instances of abnormal flows, with normal static scans. We can perhaps begin to generalize this by means of two observations.

   a. If a tumor has either increased or decreased flow as compared with surrounding tissue, or if there is abnormally rapid or delayed flow to the tumor, then the blood flow study will be positive.

   b. Since patients are receiving the radiation exposure anyway (as part of the static scans), there is no reason to deny them the flow study.

   Because of the above reasons, it is probable that radionuclide flow studies will become an increasingly important part of the search for tumors. Siemsen and co-workers (2) have pointed out the use of the flow component in liver studies. Since the human liver has only a small portion of its blood supplied by the hepatic artery,

---

* Supported by USPHS grant CA-14969 from the National Cancer Institute, and by ET-44C from the American Cancer Society.

Table 1 Methodology of the Liver Flow Study under a Gamma Camera

1. Place patient under parallel or diverging collimator to include liver.
2. Inject 55 μCi/kg body weight intravenously.
3. In adults, wait 8 seconds, then pull Polaroids each 4 seconds. Obtain one static view for localization.
4. Place data on computer tape.
5. Normal: almost no activity in liver when spleen and kidneys have filled. Liver fills about 8 seconds later.
6. Also look for: (A) Early extrahepatic filling of abdominal tumors. (B) Deviations of aorta, increased pooling (aneurysms), or delays due to compression.

the major inflow (portal vein) does not normally occur until several seconds after the aorta has filled. Early filling means relative hypertrophy of the heptatic arterial flow, hence presumably a lesion such as a tumor (most hepatic tumors have hepatic arterial blood flow) or a liver damaged by cirrhosis. We have pointed out (3) that additional data can be obtained from flow studies, such as detecting intraabdominal tumors and abnormalities of aortic flow. The technique we presently use is summarized in Table 1. An interesting case is shown in Figure 1. Here an intrahepatic metastasis had early flow (hepatic artery). Ligation of a branch of the hepatic artery was carried out surgically, and the patient had a good response.

2. *Determination of flow extraction.* Nutrients and electrolytes are extracted by tissues from the blood and lymphatic flow. Quantification of this extraction is now coming within our means, and may assume an important role in tumor detection or following tumor growth and regression. Because of the difficulty in determining

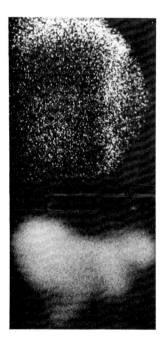

Fig. 1. Anterior liver flow and static views in a 47 year old woman with metastatic basiloid carcinoma of the anus. The flow reveals the aorta, kidney, and a band of activity above the renal mass. This band (arterial flow) corresponds to the hepatic metastasis. The patient underwent hepatic artery ligation. Case 73-409.

absolute blood flow to a tumor or other regions, we have pointed out that what can be used is an extraction ratio (4,5).

$$E = \frac{\text{Fraction of nutrient extracted in area}}{\text{Fraction of blood flow to the area}}$$

Even for deep-lying lesions, this ratio can be quantified, *if* two radionuclides with the same energy are used. By using the same energy gamma-ray emitters, the ratio cancels out the mulitple unknowns of tissue absorption and scatter. For example, $^{113m}InCl_3$ (390 keV) can be used to measure flow, while a ^{131}I nutrient (364, KeV) can be used to follow tissue extraction. There are multiple pairs of radionuclides that can be envisioned, or even two forms of the same radionuclide to measure flow and extraction.

3. *Blood pool and multiple radionuclide scans.* If a scan with one radinuclide does not give a clear-cut answer, then use of additional radionuclides is indicated. Some examples may help clarify the subject:

a. Functional asplenia is defined as the inability of the anatomically present organ to accumulate ^{99m}Tc–sulfur colloid. In such cases, blood pool scans can be helpful in showing that the spleen is present (6).

b. The "cold" thyroid nodule represents an area that is relatively lacking in the transport system for iodide or pertechnetate. A flow study, or blood pool scan, can help in determining if the region has a blood supply. This can be of use in distinguishing between a cystic structure (no flow) and a tumor (flow is present).

A further extension is shown in Figure 2. Here a cold nodule on pertechnetate scan was shown to have marked uptake of ^{75}Se selenomethionine. Thus the lesion is active metabolically. Although the radiopharmaceutical ^{75}Se selenomethionine has several disadvantages, when it is accumulated in a region, we can be assured of metabolic activity. The coupling of blood pool scanning with a ^{75}Se selenomethionine scan is shown in Figure 3. Here a lesion, noted to be devoid of ability to accumulate ^{99m}Tc–sulfur colloid, was also shown to have no blood flow and no ability to accumulate ^{75}Se selenomethionine. A cyst or necrotic lesion seems the most likely explanation.

4. *Rates of growth from scans.* Once a tumor is detected, it is important to

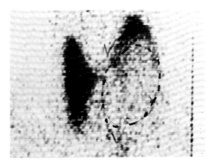 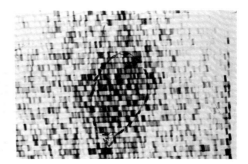

Fig. 2. Anterior rectilinear scan of the neck, performed on a 36 year old woman. The scan on the left was made after use of [^{99m}Tc]pertechnetate. A large cold nodule in the left lobe can be seen. The right-hand scan was obtained following administration of [^{75}Se]selenomethionine. The nodule has avid uptake of this radionuclide. Case 71-2282.

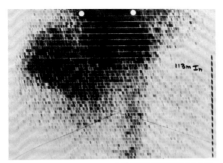
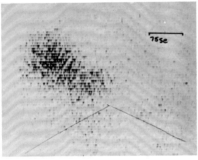

Fig. 3. Anterior rectilinear scans in an 85 year old man who had an inferior defect on a ^{99m}Tc–sulfur colloid scan of the liver. The upper view is a blood pool scan, performed with ^{113m}In. The lower scan was obtained after administration of ^{75}Se selenomethionine. The interpretation was that the area had almost no flow and no ability to accumulate the radiolabeled amino acid. Case 73-786.

know if various therapeutic modalities are causing regression. For example, intrahepatic tumors might be treated by irradiation, by chemotherapy, or by hepatic artery ligation. Scans permit an estimate to be made of organ size and tumor size. Indeed, scans are one of the few modalities that allow quantification of tumor size during life. We have just completed a study revealing that the linear growth rate of intrahepatic tumors (increase in diameter with time) (7) was on the order of 100×10^{-3} to 400×10^{-3} mm/day. This is about the same growth rate as has been reported for metastatic tumors in the lungs. Such investigations may allow differential estimation of the response of pulmonary and hepatic metastases to therapy.

INORGANIC RADIOPHARMACEUTICALS

Several inorganic radionuclides enter selected tumors following intravenous administration. This may represent an altered capillary barrier, or distorted metabolism. Unfortunately, the selection rules for this entry, and its possible value in distinguishing benign from malignant tissue is still undertermined. Some time ago, Cavalieri and co-workers (8) pointed out that ^{75}Se selenite entered some malignant tumors. Despite various studies since then (9), the long physical half-life of ^{75}Se has severely limited investigations. As will be pointed out shortly, if shorter-lived selenium radionuclides were to become available, there might be a rebirth of interest in this topic. Studies have also been reported, over a period of several years, on the use of mercurials in tumor localization. However, a recent report states that entry of this radionuclide does not distinguish malignant from nonmalignant tissue (10). Hence the use of ^{197}Hg and ^{203}Hg may be biologically severely limited (in addition to their undesirable physical characteristics). Entry of ^{99m}Tc pertechnetate into various tissues has not yet been shown to be indicative of the degree of malignancy of a lesion.

We have pursued a slightly different course, by noting that some inorganic platinum compounds have been shown to have antitumor activity. The compound *cis*-diamminedichloroplatinum (II) was synthesized in radiolabeled form, using both ^{193m}Pt and ^{195m}Pt (11,12). These investigations have shown slight localization of the radiolabel in experimental tumors, and have opened the door to a wide variety of studies.

LABELED AMINO ACIDS

Amino acids are readily transported by body tissues, and may be of potential value in following rates of uptake and metabolism. Since there is no carbon radionuclide that is a gamma-ray emitter and readily available, several stratagems have been employed for labeling amino acids.

1. Use the positron emitter ^{11}C ($T_{1/2}$ = 20 minutes) and rapid synthesis.
2. Replace one or more carbon atoms by an analog such as selenium.
3. Attach a short-lived radionuclide to the amino acid.
a. Halogenation.
b. ^{99m}Tc complexes.

These approaches are each beset by various problems.

1. *Carbon-11.* Distribution is limited by the production method, which is by means of a cyclotron or linear accelerator, ^{11}B (p,n) or other reactions. The short half-life (20 minutes) also limits distribution and incorporation into organic molecules.

2. *Selenium-75.* The long physical half-life (120 days) reduces the quantity that can be administered, because of the radiation exposure. There are three other possible selenium radioisotopes, but none is ideal: ^{72}Se ($T_{1/2}$ = 9 days); cyclotron-produced by ^{70}Ge (α,2n); the principal gamma emission is only 46 keV. ^{73}Se ($T_{1/2}$ = 7 hours); there are production difficulties, but it can be made by ^{70}Ge (α,n) or ^{75}As (d,4n); there are 359 and 511 keV emissions. ^{81m}Se ($T_{1/2}$ = 1 hour); made from enriched ^{80}Se by the reaction ^{80}Se (n,γ). There is a low yield of 103 keV photons. The ^{73}Se and ^{81m}Se might not be suitable for biosynthetic reactions because of the short half-lives, but potentially could be made into selenite, selenate, and simple organic forms.

3. *Additions to an amino acid.* (*a*) Preferably add to the side chain, so that the pH of the active groups is not changed. (*b*) Avoid attaching directly to the active sites of the amino acid.

$$\begin{array}{c} \text{COO}^- \\ | \\ \text{H}-\text{C}-\text{NH}_3^+ \\ | \\ \text{R} \end{array}$$

At pH 7.4, the zwitterion requires the ionized carboxyl and amino groups for active transport.

Labeling possibilities include: ^{123}I ($T_{1/2}$ = 13.0 hours); ^{18}F ($T_{1/2}$ = 1.7 hours); ^{99m}Tc ($T_{1/2}$ = 6.0 hours).

Several years ago the value of ^{75}Se selenomethionine in localizing lymphomas was pointed out (13,14). If this amino acid could the labeled with a shorter-lived selenium, or if other labeled amino acids were made available, the entire topic might be reopened.

The techniques for tagging pertechnetate to organic compounds are under investigation in many laboratories. One of the problems has been that the selection rules for pertechnetate labeling of molecules have never been clearly worked out. Most amino acids do not bind pertechnetate. We have found however, that reduced technetium (pertechnetate plus $SnCl_2$) can be bound to certain diamino acids. The simplest of these are:

```
        COOH                COOH
         |                   |
    H—C—NH₂              H—C—NH₂
         |                   |
    H—C—NH₂              H—C—H
         |                   |
         H               H—C—NH₂
                             |
 L-α,β-Diaminopropionic       H
         acid           L-2,4-Diaminobutyric acid
```

Further, polyamino acids made from the basic amino acids (i.e., polymers of the basic amino acids) bind reduced technetium. These include polymers of L-ornithine and of the amino acid amide L-asparagine.

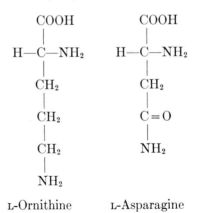

L-Ornithine   L-Asparagine

Such observations suggest that we may be on the way toward designing peptides that both bind a short-lived radionuclide and have specificity for a particular tissue. Foreign antibodies can be radiolabeled, but they themselves may trigger an antibody response. Let us look at the possibilities open to us for producing such antitissue proteins or peptides.

1. In some cases it may be possible to harvest such antibodies from patients

who have had a successful response to a foreign (tumor or other) stimulus. This *human* protein might not evoke antibody response to itself when given to a donor.

2. We might produce an antibody in an animal, which would localize in the tumor or tissue itself, and then give a human antibody to the animal protein.

3. We could potentially strip the antibody down to as small a molecular size as possible (without losing specificity), and avoid further antibody response.

4. Specific peptides might potentially be produced that have tumor or tissue specificity (and which carry a radiolabel).

RADIOLABELED ORGANIC COMPOUNDS

An area still in its infancy is the use of radiolabeled organic compounds for tumor localization. There are two principal types of compounds:

1. Radiolabeled active-site-directed enzyme inhibitors involve an interesting concept. An enzyme inhibitor, directed to an active site, carries a radiolabel. A first example of this was ^{131}I-aminopterin (15). By rate of uptake, and discharge by analogs, it may be possible to distinguish tumors from normal tissues.

2. Initial studies have been started on compounds that bind to tissue components, such as sulfhydryl groups. For example, ^{131}I- and ^{14}C iodoacetic acid have been studied (16). The hope here is that, once greater specificity is obtained, normal tissue will be distinguishable from the abnormal. Observations have also been made on diseleno compounds, which may also bind to sulfhydryl groups (17).

REFERENCES

1. Cowan, R. J., Maynard, C. D., Meschan, I., Janeway, R., and Shigeno, K.: Value of the routine use of the cerebral dynamic radioisotope study. *Radiology* **107**:111–116, 1973.
2. Waxman, A. D., Apau, R., and Siemsen, J. K.: Rapid sequential liver imaging. *J. Nucl. Med.* **13**:522–524, 1972.
3. Spencer, R. P.: Additional data on hepatic flow studies. *J. Nucl. Med.* **14**:250, 1973.
4. Spencer, R. P. and Cornelius, E. A.: Radionuclide studies of blood flow in a transplantable mouse mammary carcinoma. *J. Nucl. Med.* **12**:465, 1971.
5. Spencer, R. P. and Cornelius, E. A.: Tumor and organ uptake of nutrient: Relationship to blood flow. *Fed. Proc.* **28**:829, 1969.
6. Spencer, R. P., Pearson, H. A., and Binder, H. A.: Identification of cases of "acquired" functional asplenia. *J. Nucl. Med.* **11**:763–766, 1970.
7. Spencer, R. P. and Witek, J. T.: Radionuclide studies on the growth of intrahepatic tumors and of the infiltrated liver. *Cancer* **32**:838–842, October 1973.
8. Cavalieri, R. R., Scott, K. G., and Sairenji, E.: Selenite (^{75}Se) as a tumor localizing agent in man. *J. Nucl. Med.* **7**:197–208, 1966.
9. Jereb, M., Jereb, B., and Unge, G.: Radionuclear selenite (^{75}Se) for scinitigraphic demonstration of lung cancer and metastases in the mediastinum. *Scand. J. Respir. Dis.* **53**:331–337, 1972.
10. Gotta, H. Chwojnik, A., Seeber, J. M., and Pecornini, V.: Valoracion de la centellografia pulmonar con mercuriales. *Medicina* **32**:419–427, 1972.
11. Lange, R. C., Spencer, R. P., Harder, H. C.: Synthesis and distribution of a radiolabeled antitumor agent: *cis*-Diamminedichloroplatinum(II). *J. Nucl. Med.* **13**:328–330, 1972.

12. Lange, R. C., Spencer, R. P., and Harder, H. C.: The antitumor agent *cis*-Pt(NH$_3$)$_2$Cl$_2$: Distribution studies and dose calculation for ^{193m}Pt and ^{195m}Pt. *J. Nucl. Med.* **14**:191–195, 1973.
13. Herrera, N. E., Gonzalez, R., Schwarz, R. D., Diggs, A. M., and Belsky, J.: ^{75}Se-selenomethionine as a diagnostic agent in maligant lymphoma. *J. Nucl. Med.* **6**:792–804, 1965.
14. Spencer, R. P., Montana, G., Scanlon, G. T. and Evans, O. T: Uptake of selenomethionine by mouse and human lymphomas, with observations on selenite and selenate. *J. Nucl. Med.* **8**:197–208, 1967.
15. Johns, D. G., Spencer, R. P., Chang, P. K., and Bertino, J. R.: ^{131}I-iodoaminopterin: A gamma-labeled active-site directed enzyme inhibitor. *J. Nucl. Med.* **9**:530–536, 1968.
16. Brody, K. R., and Spencer, R. P.: Distribution and decarboxylation of 1-^{14}C-iodoacetic acid in dogs. *Int. J. Appl. Radiat. Isot.* **23**:390–392, 1972.
17. Brody, K. R., Treves, S., and Spencer, R. P.: Renal excretion of ^{75}Se-diselenodi-*N*-valeric acid. *Int. J. Appl. Radiat. Isot.* **21**:38–40, 1970.

Adrenal Tumor Localization with Iodocholesterol*

CHAPTER SIXTEEN   WILLIAM H. BEIERWALTES, M.D.

The idea for our development of a radioiodinated cholesterol for adrenal scanning came from many sources (1).

In a feasibility study with ^{14}C cholesterol in dogs, a 0.6% dose per gram uptake was achieved in the adrenal cortex (2).

Figure 1 shows the structure of ^{14}C-labeled cholesterol and the structure of the ^{125}I-19-iodocholesterol that our colleague, Raymond Counsell, synthesized (3).

The tissue distribution of radioactivity was then compared with that of ^{14}C cholesterol in 14 dogs, and of ^{125}I-19-iodocholesterol in 13 dogs pretreated with 40 units of ACTH gel daily for 2–4 days and continued to termination of the study period. The study was also conducted in 16 dogs with ^{14}C cholesterol and in 16 with iodocholesterol without ACTH (1). Figure 2 is a bar graph showing that the uptake of radioactivity in the adrenal cortex of the dog was 0.3–0.4% dose/g by 2 days without ACTH, and as high as 0.2% dose/g in 1 day with ACTH. Figure 3 is a bar graph showing that the adrenal cortex/liver radioactivity ratios continued to rise with time, however, reaching as high as 165 by the eighth day, with relatively high ratios achieved by the third day with ACTH. With this percent uptake of radioactivity from radioiodinated cholesterol, and with these excellent target/nontarget ratios, we succeeded in imaging the adrenal cortex in the dog (1) and the human (4).

As of April 1973, about 200 tracer doses of 0.5–2.0 mCi of ^{131}I-19-iodocholesterol have been given to humans in 0.75–30.0 mg in ethanol diluted to a 10% solution with normal saline and 0.2% polysorbate in < 2.0 ml of solution intravenously.

The compound decays only by thermal degradation, and we require a one-peak purity on chromatography of 95%, with a 17-day expiration date with a minimum of 70% of the radioactivity in one peak. The only reactions have been pain in the arms and legs in two patients following fast intravenous injection. The biological

---

* Supported in part by USPHS Training Grant CA-05134-18 and 11, USAEC Grant AT (11-1) 2031, The Elsa U. Pardee Foundation, Nuclear Medicine Research Fund, and the American Cancer Society Grant PRA-18.

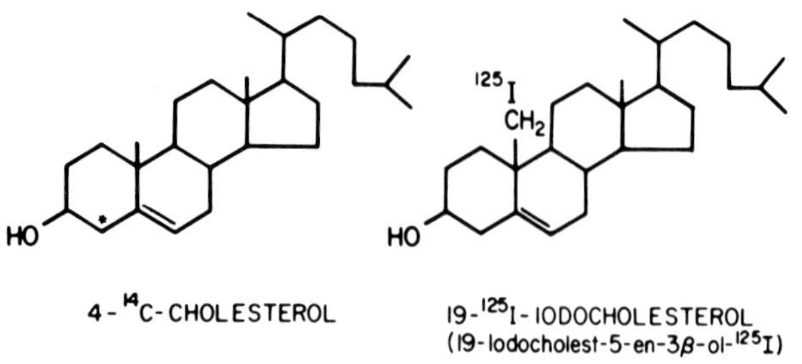

Fig. 1. Structure of ^{14}C-labeled cholesterol and the structure of the ^{125}I-19-iodocholesterol synthesized by our colleague, Raymond Counsell.

$T_{1/2}$ is 3.15 days, and the effective $T_{1/2}$ is 2.26 days, with most of the radioactivity appearing in the urine by the second day, and the peak stool excretion on the third day. The $T_{1/2}$ in blood is 2.5 hours and the adrenals receive 30–40 rads/mCi or approximately the dose of radiation delivered to the thyroid gland from a scanning dose of ^{131}I. Our latest data in humans suggests that the radiation dose to the ovaries is 1.7 rads and the testes 0.4 rads.

We have succeeded in imaging the adrenals in all patients with normal adrenocortical function (35 patients), and the percent uptake (5) at the usual scanning time of 6–10 days has ranged from 0.1 to 0.44%.

Figure 4 is a scintiphoto of the normal adrenal glands of a patient who had a resection of the right kidney for a renal carcinoma. Preoperatively, it was de-

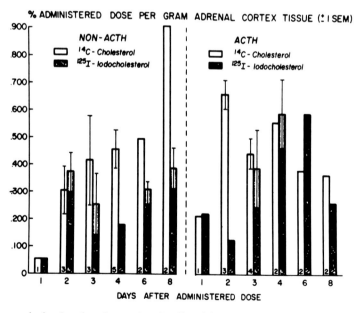

Fig. 2. Bar graph showing that the uptake of radioactivity from radiolabeled cholesterol in the dog adrenal cortex was 0.3–0.4% dose per gram by 2 days without ACTH, and as high as 0.2% dose per gram in 1 day with ACTH.

ADRENAL TUMOR LOCALIZATION

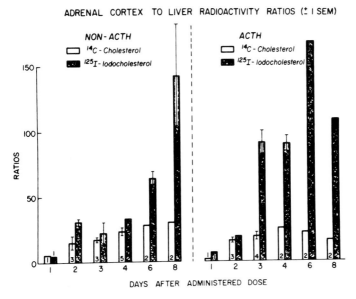

Fig. 3. Bar graph showing that the adrenal cortex/liver radioactivity ratios continued to rise with time, reaching as high as 165:1 by day 8, with high ratios after ACTH as early as day 3.

termined that he had congenital absence of the left kidney. Postoperatively, the surgeon obtained this scan to determine whether or not the patient had two adrenals and how they were functioning. Figure 5 presents scintiphotos on the left, and the computer-altered cathode ray tube display of a 4096 channel analyzer on the right showing the images of the normal adrenals in a boy with an uptake at the upper range of normal, at 10 and 14 days after the tracer.

The following table presents our experience in the differential diagnosis of adrenal tumors in relation to the presence of Cushing's syndrome.

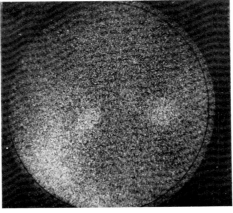

Fig. 4. Photograph of a scintiphoto of normal adrenal glands of a patient who had a percent uptake in the lower range of normal.

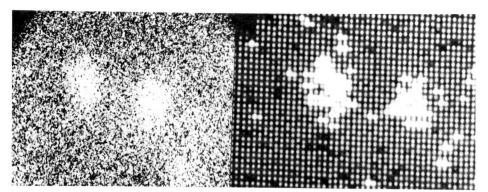

Fig. 5. Scintiphotos (left) and the computer-altered display of a 4096 channel analyzer on (right) showing bright images of the normal adrenals in a boy at 10 days.

Adrenal Tumors (April 1973)
Cushing's Syndrome

ACTH excess (bilateral hyperplasia)	7
Adrenocortical adenomas (hyperfunctioning)	7
Remnants after bilateral adrenalectomy; no Cushing's Syndrome	7
Hot nodules with and without suppression; suppressible and nonsuppressible	10

Figure 6 presents the scintiphoto adrenal images with ^{131}I-19-iodocholesterol and a rectilinear scanner superimposed upon a kidney scan performed with ^{197}Hg chlormerodrin (4). It was thought that this patient had pituitary ACTH Cushing's syndrome rather than a unilateral hyperfunctioning adrenocortical adenoma, but our venographer could not catheterize the right adrenal vein. This scan confirmed the clinical diagnosis of our endocrinologist, and he treated this patient successfully with x-irradiation to the pituitary. Figure 7 shows the typical "bright

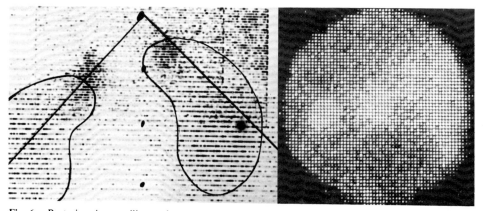

Fig. 6. Posterior-view rectilinear photoscan of our first patient's adrenal area. Both adrenal glands are imaged normally with ^{131}I-19-iodocholesterol, and this scan is superimposed on a kidney scan performed with ^{197}Hg chlormerodin (Right) Anger camera scintiphotos of same area, aided by PDP8 50/50 computer-altered display.

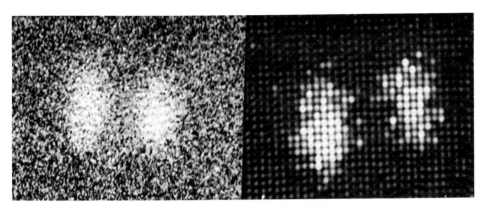

Fig. 7. Scintiphoto of the adrenals of a second patient with ACTH Cushing's syndrome, showing typical "bright headlights burning through a fog."

headlights burning through a fog" scintiphoto display of the adrenals of a second patient with Cushing's syndrome.

Figure 8 shows that, in our first group with Cushing's syndrome, all patients had higher percent uptakes than all patients without Cushing's syndrome (5).

Figure 9 shows the uptake in a hyperfunctioning adrenocortical adenoma of the right adrenal of a girl with Cushing's syndrome (left), and a venogram (right) visualizing the same tumor and showing the left adrenal gland to be atrophic. Figure 10 is a photograph of the tumor. The diameter of this tumor, as measured on the rectinlinear scan (3.5 × 3.0 cm), agreed remarkably well with the size of the tumor as measured in the operating room (3.5 × 3.0 × 2.5 cm).

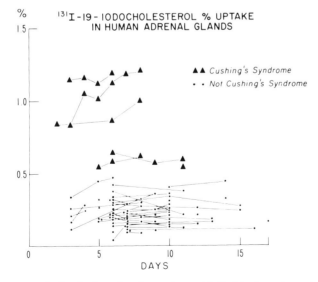

Fig. 8. The percent uptake in all of our first five patients with Cushing's syndrome had higher uptakes than all our early patients without Cushing's syndrome (5).

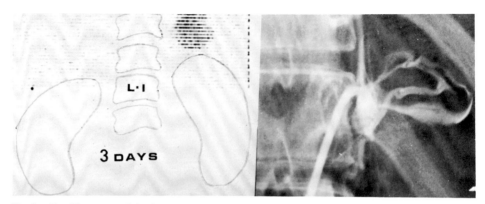

Fig. 9. Rectilinear scan (left) showing a hyperfunctioning adenoma of the right adrenal with no uptake in the rest of the right adrenal or in the left adrenal. The venogram (right) imaged the tumor and showed an atrophic adrenal on the left.

Similarly, the percent uptake measured *in vivo* (1.2% at 8 days after the tracer) agreed remarkably well with the radioactivity assay in the excised tumor (1.6% at 10 days after the tracer).

Figure 11 is the rectilinear photoscan of a right adrenal remnant, with ^{131}I Hippuran kidney localization superimposed. This patient had had bilateral adrenal gland surgery three times for Cushing's syndrome, and no adrenal tissue was found at the third operation (6).

Figure 12 is a photograph of a scintiphoto and a computer-altered display in a similar patient who was found to have four areas of radioactivity concentration in the posterior view, and a right lateral view (Figure 12b) shows the principal area of adrenal remnant. We have not yet failed to visualize the remnant in our first seven patients, and the first five have now been operated upon successfully, with the posterior and the appropriate lateral scan to localize the remnant.

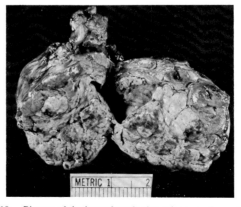

Fig. 10. Picture of the hyperfunctioning adenoma shown in Figure 9.

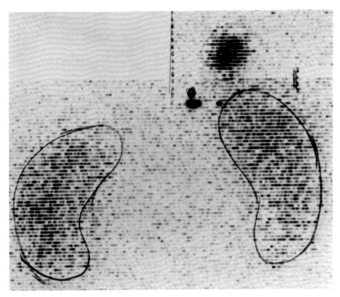

Fig. 11. Rectilinear photoscan of the right adrenal remnant with a ^{131}I Hippuran kidney localization superimposed.

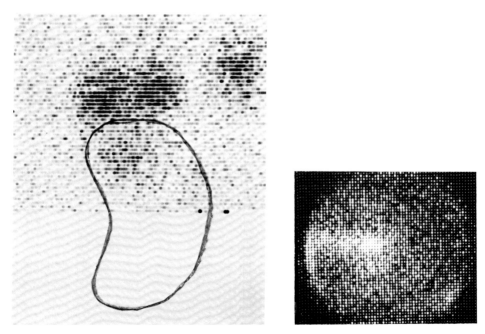

Fig. 12. (a) Photograph of the posterior view of a rectilinear scan of a patient with several right adrenal remnants. (b) Right lateral view of remnants.

The remaining tumors are as follows:

Primary Carcinoma of Adrenal Gland
With Cushing's syndrome 2
With testosterone only 1

Metastatic Carcinoma
Metastatic carcinoma to adrenal 3

Primary Adrenal Tumors
Primary aldosteronomas 14
Pheochromocytomas 6

Figure 13 shows complete lack of imaging of either adrenal gland seen in the two patterns with carcinoma of the adrenal gland producing cortisol excess. The radioactivity concentration in the first such tumor (2500 g) removed was only 0.005% dose per gram. Nevertheless, the large mass of this hypofunctioning tissue produced cortisol excess which shut off the ACTH stimulation of the normal opposite adrenal gland so that it did not concentrate ^{131}I-19-iodocholesterol.

Figure 14, however, shows that when a patient had an adrenal cortical carcinoma producing only testosterone (right adrenal gland) the testosterone did not shut off ACTH stimulation of the opposite adrenal gland to concentrate ^{131}I-19-iodocholesterol.

Figure 15 shows that bronchogenic carcinoma may replace the normal adrenal (right) so completely with metastasis that it fails to visualize. Metastases to the adrenal are seen second most commonly from breast carcinoma.

I would like to make several points about the imaging of aldosteronomas that we have observed (7,8).

1. We image aldosteronomas as a discrete tumor when they are more than 2 cm in diameter.

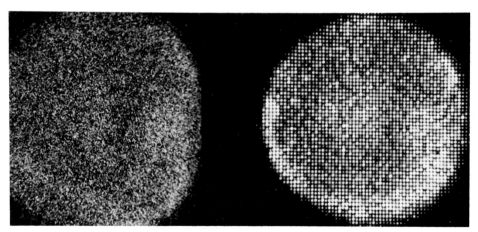

Fig. 13. Lack of imaging of either adrenal gland scan in adrenal cortical carcinoma with too little uptake in the carcinoma, but with cortical excess produced by the tumor shutting off ACTH stimulation of uptake in the contralateral normal adrenal gland.

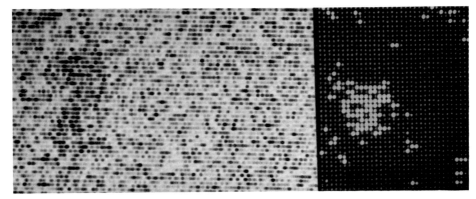

Fig. 14. Adrenocortical carcinoma producing only testosterone does not shut off uptake in the contralateral adrenal gland.

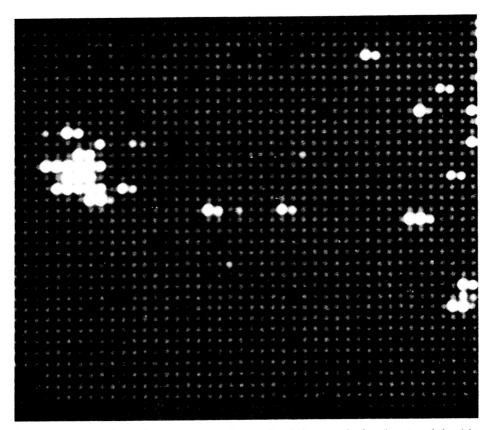

Fig. 15. Bronchogenic carcinoma metastases have replaced the normal adrenal cortex of the right adrenal gland.

2. We have lateralized the side of the aldosteronoma when the tumor is 1–2 cm in diameter by showing one adrenal image to be larger than the other.
3. We usually miss these tumors when they are less than 1 cm in diameter.
4. Dexamethasone administration usually does not suppress the uptake of ^{131}I-19-iodocholesterol in an aldosteronoma, but does suppress the uptake in a normal adrenal gland.
5. We have had one patient with idiopathic or bilateral micronodular hyperplasia whose uptake was suppressed in both adrenals with dexamethasone.
6. The uptake in the tumor does not differ significantly from the uptake in the normal adrenal.
7. The lowest concentration of ^{131}I in an aldosteronoma that permitted us to image the tumor (3 cm in diameter) was 0.03 μCi/g.

Figure 16 shows a 22 mm diameter aldosterone tumor (left) in the left adrenal, with the uptake of radioactivity unsuppressed in the remainder of that adrenal gland or in the opposite adrenal. The venogram outlined the same tumor (right).

The uptake of ^{131}I in the tumor was 0.46 μCi/g in the tumor, and 0.58 μCi/g in the normal adrenal tissue on that side.

Figure 17 shows suppression of uptake with dexamethasone in the normal left adrenal but not in a discretely imaging aldosteronoma in the right adrenal 3 cm in diameter, which concentrated 0.03 μCi/g. Figure 18 is a photograph of that tumor.

Figure 19 shows the appearance of an enlarged adrenal on the left in a patient whose tumors, in the left adrenal gland shown in Figure 20, were found to be 1 cm in diameter (below) and 6 mm in diameter (above). These tumors concentrated 0.3–0.5% dose/g while muscle contained 0.002%.

Figure 21 shows the complete suppression of uptake in both adrenals of a patient diagnosed as having an aldosteronoma on the right (the patient could not have venography because of allergy to contrast media), but proved to have bilateral micronodular hyperplasia with the right adrenal twice the diameter of the left.

Figure 22 shows that when the pheochromocytoma in this 10 year old boy reached 4 cm in largest diameter, it decreased the uptake in the right adrenal cortex sufficiently to lateralize the pheochromocytoma.

Figure 23 shows the bottom half of the right adrenal image truncated by 3 cm diameter pheochromocytoma in a 13 year old girl.

An adrenal venogram confirms the location of a tumor in 85% of aldosteronomas. The venograms have detected tumors as small as 0.5 cm. A catheter can be used to collect specimens for hormones. Venography, even when conducted by an

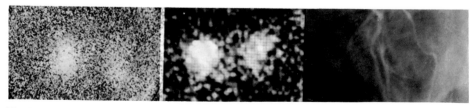

Fig. 16. Discrete imaging of a 22 mm diameter aldosterone tumor at the upper part of the left adrenal gland. Scintiphoto (left); computer-altered display (center). Venogram (right) outlines the same tumor.

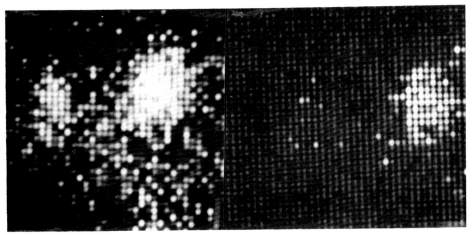

Fig. 17. Computer-altered display (left) shows discrete imaging of an aldosteronoma 3 cm in diameter in the right adrenal gland. The right panel shows suppression of uptake only in the left, and normal adrenal, after dexamethasone suppression.

expert, has several limitations, not the least of which are that the procedure is painful and requires 2 days of hospitalization. Venography cannot be used when there is sensitivity to contrast media. In 15–30% of patients, the right adrenal vein cannot be catheterized. In 5% of patients, there is rupture of medullary vessels, occasionally with infarction. Moreover, it is not certain to what degree false negatives occur. To date we have not seen a patient whose postsurgical remnant has been imaged either by arteriography or venography.

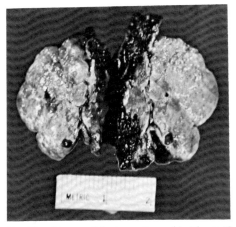

Fig. 18. Picture of the tumor imaged in Figure 17.

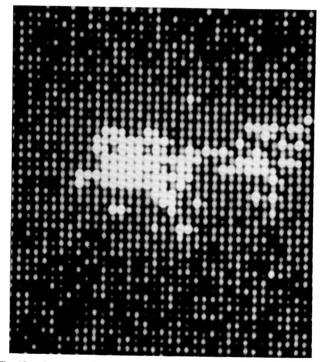

Fig. 19. Computer-altered display image of an enlarged left adrenal gland.

Fig. 20. Photograph of a cut section of the left adrenal gland shown in Figure 19. The gland contained two aldosteronomas: one 1 cm in diameter (bottom), and one 6 mm in diameter (top).

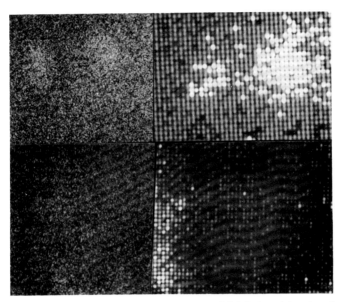

Fig. 21. Scintiphotos (left) and computer-altered display (right) before (above) and after (below), showing complete suppression of uptake in both adrenals (below) in a patient diagnosed as having a right aldosteronoma. At surgery she was found to have bilateral nodular hyperplasia (idiopathic aldosteronism), with the right adrenal twice the diameter of the left.

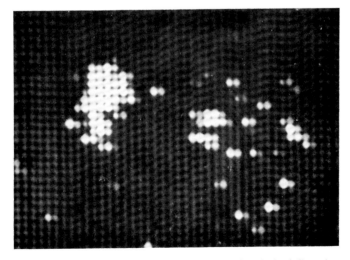

Fig. 22. Pheochromocytoma in a 10-year-old boy in the right adrenal gland distorting and distending the right adrenal cortex so that there is little or no uptake of ^{131}I-19-iodocholesterol in the adrenal cortex.

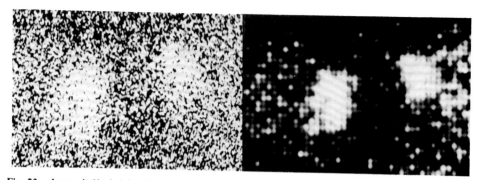

Fig. 23. Lower half of right adrenal cortex truncated by a 3 cm diameter pheochromocytoma in a 13 year old girl.

SUMMARY

1. The accuracy of adrenal iodocholesterol imaging is comparable to venography.
2. Resolution is more than 1 cm.
3. It is routine procedure.
4. It is completely noninvasive and nontraumatic.
5. No hospitalization is required.
6. Most importantly, it provides valuable functional information.

REFERENCES

1. Blair, R. J., Beierwaltes, W. H., Leiberman, L. M., Boyd, C. M., Counsell, H. E., Weinhold, R. A., and Varma, V. M.: Radiolabeled cholesterol as an adrenal scanning agent. *J. Nucl. Med.* **12**:176, 1971.
2. Beierwaltes, W. H., Varma, V. M., Lieberman, L. M., et al.: Percent uptake of labeled cholesterol in adrenal cortex. *J. Nucl. Med.* **10**:387, 1969.
3. Counsell, R. E., Ranade, V. V., Blair, R. V., Beierwaltes, W. H., and Weinfold, P. A.: Tumor localizing agents. IX. Radioiodinated cholesterol. *Steroids* **16**:317, 1970.
4. Beierwaltes, W. H., Lieberman, L. M., Ansari, A. N., Nishiyama, H.: Visualization of human adrenal glands *in vivo* by scintillation scanning. *J. Amer. Med. Assoc.* **216**:275, 1971.
5. Morita, R., Lieberman, L. M., Beierwaltes, W. H., Conn, J. W., Ansari, A. N., and Nishiyama, H.: Percent uptake of I-131 radioactivity in the adrenal from radioiodinated cholesterol. *J. Clin. Endocrinol. Metabol.* **34**:36, 1972.
6. Lieberman, L. M., Beierwaltes, W. H., Conn, J. W., Ansari, A. N., and Nishiyama, H.: Diagnosis of adrenal disease by visualization of human adrenal glands with I-131-19-Iodocholesterol, *N. Engl. J. Med.* **285**:1387, 1971.
7. Conn. J. W., Beierwaltes, W. H., Lieberman, L. M., Ansari, A. N., Cohen, E. L., Bookstein, J. J., and Herwig, K. R.: Primary aldosteronism preoperative tumor visualization by scintillation scanning. *J. Clin. Endocrinol. Metabol.* **33**:713, 1971.
8. Conn, J. W., Morita, R., Cohen, E. L., Beierwaltes, W. H., MacDonald, W. J., and Herwig, K. R.: Primary aldosteronism: Photoscanning of tumors after I-131-19-iodocholesterol, *Arch. Intern. Med.* **129**:417, 1972.

The Present Status of the ^{32}P Test in Ophthalmology

CHAPTER SEVENTEEN

PAUL L. CARMICHAEL, M.D.
GERALD C. HOLST, Ph.D.
JAY L. FEDERMAN, M.D.
JERRY A. SHIELDS, M.D.

INTRODUCTION

The need for accurate identification of intraocular malignant melanoma in ophthalmology has been stated by numerous authors. Ferry (1), in his study from the Armed Forces Institute of Pathology (AFIP), stated that the incidence of *misdiagnosis* of malignant melanoma ran as high as 19%. This figure was disputed by others (2,3), who have reported statistics between 6 and 8%. Recently, however, Shields and Zimmerman (4) in a review of over 5000 cases from the AFIP files, have again confirmed the incidence of misdiagnosis as approximately 20%. It is felt that an error of 20% is typical of all hospitals and is not limited to teaching centers (Table I).

Several lesions may stimulate malignant melanoma of the uvea. Among these are hemangioma, nevus of the choroid, retinal detachment, metastatic carcinoma, and choroidal hemorrhage or hematome (Table 2). Since malignant melanoma is a potentially fatal lesion it is imperative that testing be undertaken with various modalities to improve the diagnostic acumen of the clinician.

With the invention of the binocular indirect ophthalmoscope, a new approach to intraocular tumor localization was introduced to ophthalmologists. Tests such as ultrasonography (5) and transillumination are good techniques for differentiating between solid and serous detachments. Fluorescein and angiography has also been recommended as a potential method for tumor identification (6). Unfortunately, fluorescence around the tumor may be blocked by heavy pigmentation or subretinal hemorrhage which may surround the tumor site (7).

Table 1 Pseudomelaonmas of the Posterior Uvea

Study	Years	Number of Enucleations	Number of Diagnoses of Malignant Melanoma (Clear Media)	Incorrect Diagnoses	Percent
Ferry	1957–1962	7877	529	100	19
Blodi and Roy	1940–1966	1375	95	6	5.6
Howard	1957–1965	—	214	17	8
Shields and Zimmerman	1963–1970	5889	208	41	20
Shields and McDonald	1962–1972	1398	188	7	3.7

HISTORY

The radioactive phosphorous (^{32}P) uptake test was first introduced to opthalmology by Thomas in 1952 (8). Previous work by Low-Beer and (9) Selverstone (10) had laid the groundwork for this development. Radiophosphorous was felt to be useful for identification of malignant tumors of the choroid, since these tumors seemed to concentrate ^{32}P in their DNA, RNA, acid-soluble, and pholsphold fractions (11). Early difficulties in differentiating between vascular, inflammatory, and malignant lesions were overcome by testing at 1 and 24 hours after injection (12). The test was found by several investigators to be almost 95% reliable in accessible lesions. Nevertheless, a serious limitation remained, inasmuch as posteriorly located lesions

Table 2 Lesions Enucleated Suspected Uveal Melanomas

	Ferry (%)	Shields, Zimmerman (%)
Rhegmatogenous retinal detachment	29	34
Disciform macular degeneration	9	10
Chorio retinitis	8	7
Senile retinoschisis	7	7
Lesions of retinal pigment epithelium	7	5
Choroidal detachment	7	10
Choroidal hemangioma	5	10
Metastatic carcinoma	5	7
Choroidal nevus	5	0
Melanocytoma of optic nerve	4	2
Vitreous hemorrhagic	4	2
Hemorrhagic detachment of macula	3	10
Others	8	2

were inaccessible despite the availability of a "posterior" probe (13–15) since localization of the tumor with a direct opthalmoscope was extremely difficult. Although introduction of the binocular indirect ophthalmoscope solved the problem of localization, the suggestion by Dunphy that it be used (29) lay dormant from 1964 until 1970 when Hagler and Jarrett (30) reintroduced the test as a highly accurate method for detection of intraocular malignant melanoma utilizing the 60% 48 hour criteria of Goldberg and Kara (28).

From the earliest experiences with the test, it has proved over 95% accurate in its detection capabilities for malignant melanoma (16–29). The statistics continue to be repeated by centers in which the test is continually performed (32–34).

Further refinement of the test was given impetus by the introduction of a semiconductor detector by LaRose (31) and co-workers. The instrument consists of a lithium-drifted silicon probe which may be almost one order of magnitude more sensitive to beta energies than a Geiger-Müller detector. The higher sensitivity tends to permit better counting statistics for iris lesions and lesions smaller than 3 mm in diameter.

MATERIALS AND METHODS

At present there are two commercially available instruments for performing the tumor identification test. One is a Geiger-Müller probe,* and the other a lithium-drifted silicon probe.† Both probes are designed to pass posteriorly, and are connected to suitable rate meters designed expressly for each instrument. The Geiger probe operates at 550–650 V, whereas the output voltage of the semiconductor detector is 5 V. The silicon probe has a threshold of 130 keV for beta energies, whereas the Geiger probe operates at a much higher threshold.

The test as presently performed consists of an intravenous injection of 10 μCi/kg ^{32}P, and counting is perfomed 48 hours later. If the lesion is situated in the choroid or ciliary body area anterior to the equator, the counts are made directly over the conjunctiva in the area of the tumor site, and the opposite eye is used as a control.

If the lesion is situated posteriorly, the conjunctiva is cut and the tumor accurately localized with an indirect ophthalmoscope. After localization, diathermy marks are placed in the area surrounding the tumor, and the posterior probe is then placed over the site of the tumor. A series of four to five 1 minute readings is performed over the tumor. A quadrant 180° from the tumor site is used as a control area. The test may be performed under local anesthesia, but is done more often under general anesthesia since enucleation of the globe usually follows a positive test. Accurate placement of the probe is essential, since a drift of 1–2 mm from the tumor site may result in fewer counts and a false negative result.

RESULTS

From February 1970 to February 1972, 121 cases of suspected intraocular malignant melanoma were studied at the Wills Eye Hospital with the ^{32}P test. Of these cases 50 had histological confirmation. There was only one false negative test in the

---

* EON Corporation, Brooklyn, N.Y.
† Nuclear Associates, Westbury, N.Y.

entire series, giving the test a better than 95% accuracy for detection of intraocular melanomas. The distribution of these cases is seen in Tables 3 and 4.

Intuitively it seems that the size of a lesion might influence the number of counts generated during a ^{32}P test. Although this has not been our experience in all cases, mathematical calculations were performed* to evaluate a possible relationship between tumor size and the number of counts recorded during a typical study.

Figure 1 shows the energy spectrum of ^{32}P in air. Using the mass attenuation coefficient for water, the energy spectrum can be converted to a "tissue penetration" spectrum. The probability that an electron will travel (penetrate) a distance x is given by the probability density function $p(x)$ (Figure 2). The fraction of the total number of electrons that penetrate a distance of at least y is given by

$$\int_{y}^{8.5} p(x)\, dx$$

where the maximum penetration distance is 8.5 mm. To calculate the fraction of electrons from the tumor that reach the detector, a simple one-dimensional model is considered in which the following basic assumptions are made:

1. The sclera is 1 mm thick.
2. The detector element is recessed 1 mm within the probe.
3. The choroidal coat is 0.4 mm thick.
4. The base of the tumor sits within the choroid.
5. The sclera does not absorb an appreciable amount of ^{32}P.

The distances involved are schematically shown in Figure 3.

For purposes of calculation, the choroid and tumor are divided in 0.1 mm layers, and the contribution from all layers is summed to find the total fraction of electrons reaching the detector. For each layer it is assumed that the penetration spectrum is the same, but the relative concentration of ^{32}P may vary. Let N be the concentration of ^{32}P in a layer of normal tissue, and let M be the concentration in a

Table 3 ^{32}P Studies Performed

Location	Number of patients
Iris	18
Anterior choroid	30
Posterior choroid	73
Total	121

* The mathematical model and calculations were prepared by G. Holst, consulting physicist, Retina Service, Wills Eye Hospital.

RESULTS

Table 4

Location of Lesion	Tissue Specimens	Pathology			Correct Diagnosis with ^{32}P Test
		Melanoma	Metastatic	Benign	
Iris	2	1	0	1	2
Anterior choroid	11	10	1	0	11
Posterior choroid	37	35[a]	2	0	36

[a] One false negative test.

layer of abnormal tissue. Then, for normal choroid the total fraction reaching the detector is:

$$\int_{2.0}^{8.5} Np(x)\, dx + \int_{2.1}^{8.5} Np(x)\, dx + \cdots + \int_{2.3}^{8.5} Np(x)\, dx = 1.553N$$

All tumor cells beyond 6.1 mm from the base of the tumor do not contribute to the number of electrons counted, because these cells are more than 8.5 mm from the detector. The fraction reaching the detector for an "infinite" tumor is

$$\int_{2.4}^{8.5} Mp(x)\, dx + \int_{2.5}^{8.5} Mp(x)\, dx + \cdots + \int_{8.4}^{8.5} Mp(x)\, dx = 7.207M$$

Using the criterion that the rate of uptake in abnormal tissue must exceed 50%, the following ratio holds:

$$\text{Ratio} = \frac{(\text{tumor counts} + \text{choroid counts}) - \text{choroid counts}}{\text{Choroid counts}} \geq \tfrac{1}{2}$$

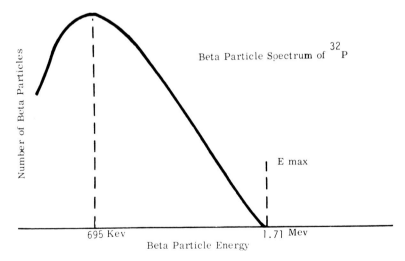

Fig. 1. The energy spectrum of ^{32}P in air.

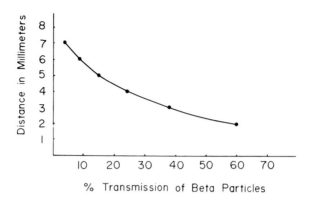

Fig. 2. A "tissue penetration" spectrum for ^{32}P. This is derived from the energy spectrum by the probability tensity function $p(x)$.

or

$$\text{Tumor counts} \geq \tfrac{1}{2} \text{ choroid counts}$$

from above

$$7.207M \geq \tfrac{1}{2}[1.553N]$$

$$M \geq 0.11N$$

Thus to reach the criterion of 50%, the tumor tissue need only to adsorb one-ninth as much ^{32}P as normal tissue. Experimentally (11), it is known that ^{32}P must enter vitreous at the same rate as it is absorbed in the choroid. Then the counts over normal tissue are due to

$$\text{Choroid counts} + \text{vitreous counts}$$

and the counts over the tumor are due to

$$\text{Choroid counts} + \text{tumor counts} + \text{remaining vitreous counts}$$

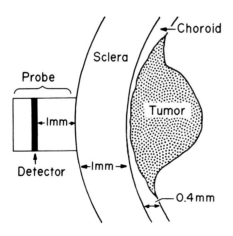

Fig. 3. Schematic representation of distance within the eye and probe system.

RESULTS

Table 5

Tumor Thickness	Tumor Counts	Remaining Vitreous Counts
1 mm	3.142 M	4.065 N
2 mm	5.262 M	1.945 N
3 mm	6.487 M	0.720 N
4 mm	7.032 M	0.175 N
4 mm	7.187 M	0.020 N
Infinite	7.207 M	0.000

the uptake ratio becomes

$$\frac{\text{Tumor counts} + \text{remaining vitreous counts} - \text{normal vitreous counts}}{\text{Choroid counts} + \text{vitreous counts}}$$

where

$$\text{choroid counts} = \int_{2.0}^{8.5} Np(x)\,dx + \cdots + \int_{2.3}^{8.5} Np(x)\,dx = 1.553N$$

$$\text{normal vitreous counts} = \int_{2.4}^{8.5} Np(x)\,dx + \cdots + \int_{8.4}^{8.5} Np(x)\,dx = 7.207N$$

$$\text{tumor thickness } X = \int_{2.4}^{8.5} Mp(x)\,dx + \cdots + \int_{X-1}^{8.5} Mp(x)\,dx$$

$$\text{remaining vitreous} = \int_{X}^{8.5} Np(x)\,dx + \cdots + \int_{8.4}^{8.5} Np(x)\,dx$$

The results are listed in Table 5.

Table 6

Size of Lesion	M/N
1 mm	2.40
2 mm	1.83
3 mm	1.67
4 mm	1.62
5 mm	1.61
Infinite	1.61

Table 6 lists the results using a 50% uptake criterion.

The fact that M/N is always greater than 1 means that the tumor has absorbed ^{32}P at a higher rate than normal tissue. Since M/N is nearly constant for lesions over 3mm in thickness, the calculations indicate that with present tissue detector geometry there is probably no relationship between tumor size and rate of uptake of ^{32}P for tumors more than 3 mm in thickness.

COMMENTS

Our studies agree with Hagler and Jarrett (33) that the ^{32}P test continues to be the most accurate test devised for the diagnosis of malignant melanoma.

The one false negative test was one in which the probe could not be accurately placed because the lesion was pedunculated and adjacent to the optic disc. There were no false positive tests in this series.

An attempt to correlate the percentage uptake with the cell type of the melanoma has been unsuccessful to date. Perhaps further studies may elucidate what should be a possible connection between the metabolic activity of a given tumor and amount of radiophosphorus uptake within the lesion. At present we are studying the relative quantities and localizations of ^{32}P in normal choroid, retina, cornea, lens, tumor, and vitreous. Our studies at this time lead us to believe that a significant amount of ^{32}P enters the vitreous cavity around the tumor. This knowledge should lead to a more precise understanding of the true basis for the clinical test as it is now performed.

From Table 6 the uptake of ^{32}P in tumor over normal tissue must increase as lesion size decreases to maintain the 50% criterion. However, it can be assumed that the M/N ratio is always fixed, regardless of tumor size. Therefore in order to maintain a constant M/N the criterion must increase manyfold as tumor size decreases below 3 mm. This may be the reason why the test has been so successful with the Geiger-Müller counter in the past, and possibly why false negative tests seem to outnumber false positive tests almost 2 to 1 in the literature. Interpretation of tests from lesions smaller than 3 mm in size and lesions of greater depth or with eccentric areas of activity may lead to false negative tests unless a very sensitive detector is utilized. The higher sensitivity of the silicon probe should make it ideal in those instances. Avalanche semiconductors have also been utilized as highly sensitive detectors for diagnosis of ocular malignant melanoma. However, the probes were found to be bulky and too difficult to place on posteriorly located tumors. The low counting rate and a need for a relatively dry environment and constant temperature were found to be further problems affecting their clinical usefulness (35).

It seems that the future reliability and advantage of any tumor detection test will lie in its ability to detect activity in small tumors previously described as nonactive and nonprogressive by clinical observation. If this advantage can be capitalized upon by the new semiconductors, modified avalanche detectors, or noncontact detector imaging systems, small lesions could be identified as malignant at an earlier stage and might be treated by photocoagulation, cryosurgery, radiation, or surgical excision before they require enucleation of an otherwise healthy globe.

REFERENCES

1. Ferry, A. P.: Lesions mistaken for malignant melanoma of the posterior uvea. *Arch. Ophtha* 7. Pettit, T. H. Barton, A. Foos, R. Y., and Christensen, R. E. Fluoroscein angiography of choroidal melanomas. *Arch. Ophthalmol.* **83**:27–38, 1970.
8. Thomas, C. I., Krohner, J. S., and Storaasli, J. P.: Detection of intraocular tumors with radioactive phosphorous, *Arch. Ophthalmol.* **47**:276, 1952.
9. Low-Beer, B. V. A.: Surface measurements of radioactive phosphorous in breast tumors as a possible diagnostic method. *Science* **104**:399, 1946.
10. Selverstone, B., Solomon, A. K., and Sweet, W. H.: Localization of brain tumors by means of radioactive phosphorous, *J. Amer. Med. Assoc.* **140**:277, 1946.
11. Thomas, C. I., Harrington, H., and Bovington, M. D.: Uptake of radioactive phosphorous in experimental tumors. III. Biochemical fate of ^{32}P in normal and neoplastic ocular tissue. *Cancer Res.* **18**:1008, 1958.
12. Terner, I. S., Leopold, I. H., and Eisenberg, I. J.: The radioactive phosphorous uptake test in ophthalmology. *Arch. Ophthalmol.* **55**:52, 1956.
13. Thomas, C. I., Kroher, J. S., and Storaasli, J. P.: Geiger counter probe for diagnosis and localization of posterior intraocular tumors. *Arch. Ophthmol.* **52**:413–414, 1954.
14. Bettman, J. W. and Fellows, V.: Radioactive phosphorous as a diagnostic aid in ophthalmology. *Arch. Ophthalmol.* **51**:171, 1954.
15. Dunphy, E. B.: Experience with radioactive phosphorus in tumor detection *Trans. Amer. Ophthalmol. Soc.*, **54**:289, 1956.
16. Thomas, C. I., Krohmer, J. S., Storaasli, J. P.: Detection of intraocular tumors by use of radioactive phosphorous. *Am. J. Ophthamol.* **38**:93, 1954.
17. Kennedy, R. J., Glosser, O., and Kozden, P.: The use of radioactive phosphorous in detection of intraocular tumors. *Cleveland Clin. Quart.* **21**:133, 1954.
18. Palin, A. and Tudway, R. D.: The uses of radioactive isotopes in the diagnosis of intraocular malignancy, *Trans. Ophthalmol. Soc. U.K.* **55**:281, 1955.
19. Donn, A. and McTigue, J. W. The radioactive uptake test for malignant melanoma of the eye. *Arch. Ophthalmol.* **57**:668, 1957.
20. Dunphy, E. B., Dowling, J. L., Jr., and Scott, N. A.: Experience with radioactive phosphorous in tumor detection. *Arch. Ophthalmol.* **57**:485, 1957.
21. Shapiro, I. Radioactive phosphorous in differential diagnosis of ocular tumors. *Arch. Ophthalmol.* **57**:14, 1957.
22. Thompson, G. A. The use of radioactive phosphorous in the diagnosis of ocular tumors. *Trans. Can. Ophthalmol. Soc.* **10**:98, 1958.
23. Dollfuss, M. A., Guerin, R. A., and Guerin, M.D.: Radioisotope and ocular tumors *Arch. Ophthalmol.* **18**:5, 1958.
24. Perkins, E. S. and Duguid, I.: P_{-32} and tumors of the eye. XVIII. Concilium. *Ophthalmol. (Belg.) Acta* **1**:506, 1958.
25. Palin, A.: Radioactive phosphorous for the detection of ocular tumors. XVIII. Concilium. *Ophthalmol (Belg.) Acta:* **1**:511, 1958.
26. Kleifeld, O. and Hochwin O.: The transmissability of radioactive phosphorus in pathological and normal human eyes. *Ber. Deutsch. Ophthalmol. Ges.* **62**:86, 1959.
27. Carmichael, P. L. and Leopold, I. H.: Radioactive phosphorus test in ophthalmology. *Amer. J. Opthalmol.* **49**:484, 1960.
28. Goldberg, A. Tabonitz, O., Kara, G. B., Zavell, A., Esperita, R.: Use of P-32 in diagnosis of intraocular tumors. *Arch. Ophthalmol.* **65**:196, 1961.
29. Leopold, I. H., Keates, E. U., Charkes, I. D.: Role of isotopes in diagnosis of intraocular neoplasm. *Trans. Amer. Opthalmol. Soc.*, **62**:89, 1964.
30. Hagler, W. S., Jarret, W. H., Humphrey, W. T.: The radioactive phosphorous uptake test in diagnosis of uveal melanoma. *Arch. Ophthalmol.* **83**:548, 1970.

31. LaRose, J. H., Jarrett, W. J., Hagler, W. S., Palms, J. M., Wood, R. E.: Medical problems in eye tumor identification *IEEE Trans. Nucl. Sci., N.S.* **18(1)**:46, 1971.
32. Carmichael, P. L., Federman, J., Shields, J., and Holst, G.: Further considerations of ^{32}P studies in ocular melanomas. Presented at a meeting of the Association for Research in Vision and Ophthalmology, May 3, 1973.
33. Hagler, W. J., Jarret, W. H., Schnauss, R. H. LaRose, J. H. Palms, J. H., and Wood, R. E.: The diagnosis of malignant melanoma of the ciliary body or choroid. *South. Med. J.* **65**:1, 1972.
34. Taylor, B., and Carmichael, P. L.: Further studies with the radioactive phosphorous test. Presented at the 23rd Wills Eye Hospital Conference, February 1972.
35. LaRose, J. H.: Semiconductor detectors for eye-tumor diagnosis. In Hoffer, P. B., Beck, R. N., and Gottshalk, A. (eds.), *Semiconductor Detectors in the Future of Nuclear Medicine.* Society of Nuclear Medicine, Inc., New York. 1971, Chapter 13.

The Characteristic Radionuclide Appearance of Certain Pediatric Central Nervous System Neoplasms

CHAPTER EIGHTEEN JAMES J. CONWAY, M.D.

INTRODUCTION

Much of what appears in this volume on radionuclide studies of neoplastic disease is just as applicable for the child as it is for the adult. In fact, it would be a formidable task to describe all the studies, techniques, radiopharmaceuticals, dosimetry, and advantages of radionuclides for the study of neoplastic disease in the pediatric patient. There are peculiarities about certain pediatric neoplasms, however, which lend a characterization to the appearance of the lesion and thus enhance their diagnostic interpretability. An illustration and discussion of these characteristics would perhaps be of benefit to those who occasionally or even commonly interpret pediatric studies in their practice. The central nervous system, in addition to being a more common site for neoplasia in the child, permits one of the more easily performed studies in pediatric nuclear medicine. Thus a large portion of current pediatric nuclear medicine practice involves the central nervous system. I therefore relegate the following discussion to a characterization of central nervous system neoplasms in the pediatric age group.

The characterization of brain lesions is dependent upon two parameters: the intensity of radionuclide accumulation within the lesion, and its location within the brain. The combination of these factors produces minimal overlap in the differentiation of the various neoplasms that occur within the central nervous system. Of course, one must always be aware, as in all areas of medicine, that individual human variability prevents an all-or-none phenomenon. The term pathognomonic" seldom if ever can be applied to diagnostic nuclear imaging.

MATERIAL AND METHOD

Experience has been drawn from approximately 2000 brain imaging studies in children during the past 5 years. The method of study has varied considerably as the evolution of new and better radiopharmaceuticals, imaging equipment, and techniques has developed. At present, sedation is employed when indicated (1), and potassium perchlorate is given routinely (2). ^{99m}Tc pertechnetate currently is the radiopharmaceutical of choice, and is administered in a dosage of 100 μCi/lb. This approximates 1 mCi/year in most children. Other radiopharmaceuticals such as ^{111}In-DTPA (3) offer the advantage of minimal accumulation in the choroid plexus and salivary glands but have not been shown to increase the diagnostic accuracy of brain imaging.

In most instances, dynamic studies in the form of a radionuclide cerebral angiogram are recorded in the posterior projection onto a videotape data storage device. These images are permanently recorded on Polaroid film at a latter interval. The perfusion characteristics of neoplasms can be visualized during the arterial and venous phases of the angiogram. As in the adult (4), there is a tendency for relatively vascular tumors to appear early in the examination. Poorly vascularized lesions such as cystic neoplasms have a tendency to appear later in the examination. "Static" brain images in the anterior, posterior, both lateral, and vertex positions are recorded with a gamma camera during the first hour following injection. Delayed images, at a minimum of 2 hours, are usually obtained with a rectilinear scanner. Delayed images of less cooperative patients or of very small infants are better studied with a gamma camera. Special views, such as oblique views (5) and pinhole magnification views (6), are frequently used to clarify the appearance of lesions, particularly in the posterior fossa. Older and more cooperative children with near-adult head sizes, almost always have delayed rectilinear scan images following the dynamic and early static gamma images. The capabilities of both imaging systems are therefore used to best advantage, for example, visualization of superficial structures on early images with a gamma camera and deeper images with focused rectilinear scans at delayed intervals.

CHARACTERIZATION OF NEOPLASMS

The characterization of neoplasms is based primarily upon two parameters:

1. Localization of the lesion.
    a. Infratentorial versus supratentorial.
    b. Midline versus central or peripheral.
    c. Anterior versus posterior.
2. The intensity of radionuclide accumulation within the lesion.
    −1 (Cold)—decreased localization.
     0 (Cool)—normal localization.
    +1 (Warm)—moderate localization.
    +2 (Hot)—intense localization.

There are several factors that complexly affect the intensity of radionuclide localization within brain neoplasms. These factors include the total mass of the lesion, the blood pool or vascularity within the abnormal tissue, the viability of the cellular elements comprising the neoplasm, and the histology of the cellular structure. Other more complex mechanisms such as the effect of corticosteroids and chemotherapeutic agents are poorly understood factors which also influence the radionuclide accumulation within neoplasms.

Supratentorial Midline Lesions

Radioactive		Nonradioactive
2+ (Hot)	1+ (Warm)	0 (Cool)
Optic glioma		Craniopharyngioma
Glioblastoma multiforme		Teratoma
		Pinealoma
	Ependymoma	

The most common pediatric lesions encountered in this group are optic glioma and craniopharyngioma. The intense localization of ^{99m}Tc pertechnetate in optic glioma has been a characteristic finding. These lesions often attain a considerable size before recognition of symptomotology by the parent or clinician. The large bulk of abnormal tissue probably contributes to the intensity of radionuclide accumulation. Radionucleic images (Figure 1) combined with roentgenographic changes in sella turcia or optic foramen enlargement in an infant or young child with optic tract signs confirms the diagnosis.

Craniopharyngioma, conversely, is found in the older child. Calcifications are often noted roentgenographically within this neoplasm, and the sella turcica is often enlarged. Radionuclide localization, however, is often very poor in spite of frequently bulky tumors. Craniopharyngiomas may also be cystic, which may further contribute to the lack of radionuclide accumulation with these lesions. Particular attention to imagining detail (7) improve the detectability of this neoplasm.

A review of the world's literature indicates a 100% accuracy in detecting the glioblastoma multiforme neoplasm. This lesion usually presents with an intense radionuclide localization, but may be located in almost any area of the brain. It

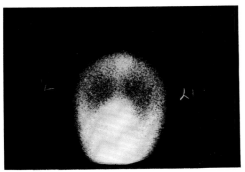 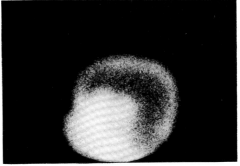

Fig. 1. The midline (*a*) and suprasellar (*b*) location of the lesion, combined with an intense radionuclide localization, characterizes this lesion as an optic glioma. The diagnosis is further certified by the clinical presentation, that is, optic tract signs in an infant or young child with an enlarged sella turcica or optic foramina.

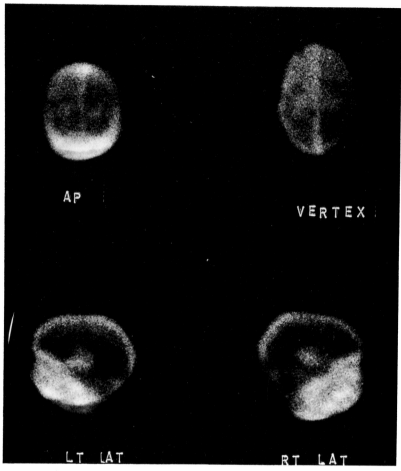

Fig. 2. The lesion crosses the midline as seen in the anterior and vertex views. Such an appearance indicates a highly infiltrative lesion crossing the midline through a commissure, and is characteristic of glioblastoma multiforme.

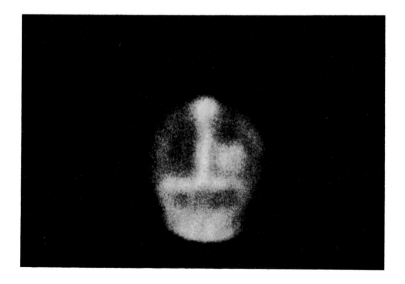

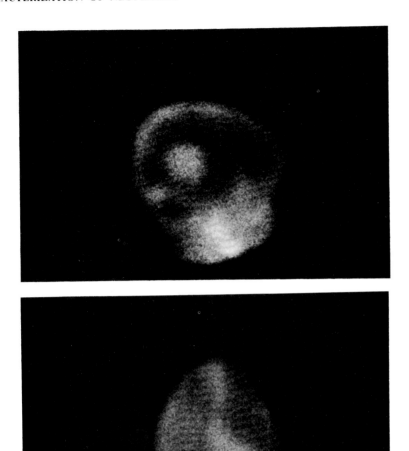

Fig. 3. An intense localization of radionuclide in the midposterior aspect of the cerebrum is characteristic of the choroid plexus papilloma. Potassium perchlorate has little effect upon the accumulation of ^{99m}Tc by this neoplasm. Hydrocephalus, which is predominantly on the side of the lesion, substantiates the diagnosis.

therefore cannot be excluded from the differential diagnosis of those lesions with intense radionuclide localization. Most commonly they are located within the cerebral hemispheres and not within the midline or in the periphery. A most useful radionuclide sign is the presentation of the lesion on both sides of the midline (Figure 2). Such an appearance suggests the probability of a highly infiltrative lesion such as a glioblastoma multiforme that has crossed the midline through a commissure.

Finally, one must always consider teratoma, pinealoma, and ependymoma when dealing with midline lesions. Calcifications are frequently present in teratoma and pinealoma. Teratoma is usually well differentiated and localizes radionuclide poorly. Ependymoma can arise anywhere within the ventricular system, and therefore can present in the midline third and fourth ventricles. Pinealoma is located posteriorly in the region normally occupied by the pineal gland.

Supratentorial Cerebral Lesions

2+ (Hot)	1+ (Warm)	0 (Cool)
Choroid plexus papilloma Glioblastoma multiforme	Astrocytoma grade I–II	

Choroid plexus papilloma has a rather characteristic appearance when it originates in its most common locus at the posterior aspect of the lateral ventricles (Figure 3). Localization of radionuclide within these lesions is intense in spite of adequate administration of potassium perchlorate (8). In one instance, brain imaging with ^{99m}Tc pertechnetate was performed with and without potassium perchlorate, and no apparent difference in radionuclide localization was noted. One must remember that choroid plexus tissue can be found throughout the ventricular chambers, and thus may present as a midline lesion within the third and fourth ventricle (9,10).

Astrocytoma grade I-II becomes an enigma in brain imaging, for on occasion even large tumors fail to visualize. Perhaps this is due to the well-differentiated nature of some of these lesions. Occasionally, they are even cystic. Generally, their lesser degree of radionuclide accumulation readily distinguishes them from glioblastoma multiforme.

Metastatic disease, common in the adult, is most unusual in the child. Indeed, a metastic lesion has not been noted in the material at the Children's Memorial Hospital during the past 5 years. Whenever multiple loci of radionuclide accumulation have been found, they have occurred within multiple abscesses, infarctions, or inflammatory processes.

Supratentorial Peripheral Lesion

2+ (Hot)	1+ (Warm)	0 (Cool)
Meningeal sarcoma		

Probably the most common peripheral supratentorial lesion in the adult is meningioma. Meningioma in childhood is uncommon. Its counterpart, meningeal sarcoma presents more frequently and is characteristic in appearance. An intense localization of the radionuclide in the periphery of the cerebrum labels this tumor with considerable certainty (Figure 4).

Infratentorial Midline Lesions

2+ (Hot)	1+ (Warm)	0 (Cool)
Medulloblastoma Ependymoma	Teratoma	Brain stem glioma

The most common pediatric midline lesions occurring in the posterior fossa are medulloblastoma and brainstem glioma. These are easily differentiated, since medulloblastoma has a moderately intense degree of radionuclide localization, whereas brainstem glioma generally is poorly visualized. Earlier reports on the detection of posterior fossa lesions in childhood were less than optimistic. This was due to inadequate technique in visualizing the posterior fossa, and perhaps also to the radiopharmaceuticals that were used. More recent reports indicate a very high degree of accuracy for detecting medulloblastoma (10). A bulky tumor may extend laterally, and thus prove difficult in differentiating from cerebellar astrocytoma.

Brainstem glioma in childhood remains the most difficult tumor to detect, with only about 40% of such neoplasms recognized by brain imaging. This occurs in spite

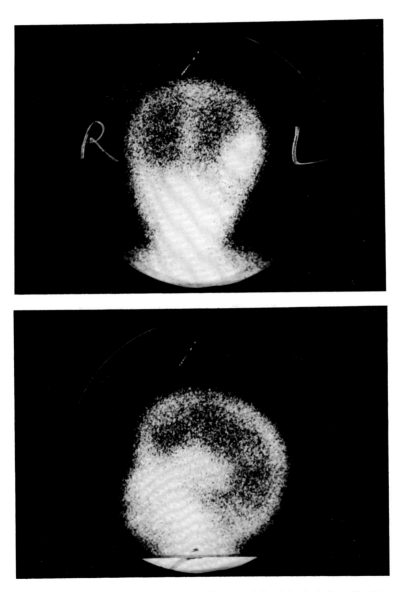

Fig. 4. Anterior (*a*) and left lateral (*b*) views localize a peripheral brain lesion. The intense accumulation of radionuclide and peripheral location are characteristic of meningeal sarcoma.

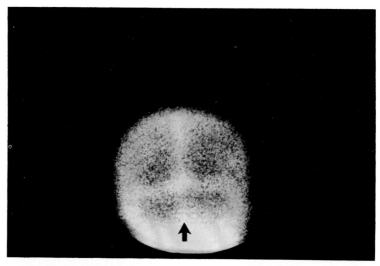
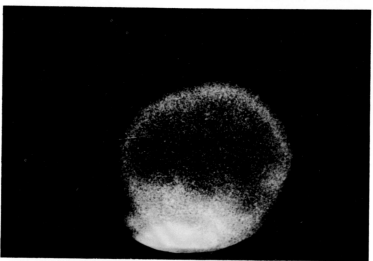
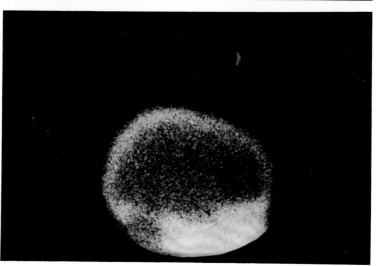

210

CHARACTERIZATION OF NEOPLASMS

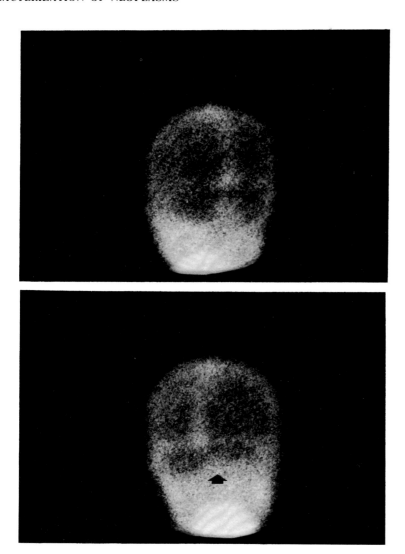

Fig. 5. The delayed posterior image (*a*) illustrates midline activity in the posterior fossa (arrow). This cannot be differentiated from an occipital venous sinus which is commonly visualized in children. The lateral views (*b* and *c*) suggest normal posterior fossa contents. The clinical presentation, strongly suggestive of a brainstem lesion, warranted special oblique views (*d* and *e*), which illustrate that the midline activity noted in the posterior view does not stay in the plane of the sagittal sinus as would be expected with an occipital venous sinus, but is deeply located in the brainstem (*e*, arrow). This is a brainstem glioma which would be difficult to diagnose on the routine views.

of large tumors which trespass into the posterior fossa or even supratentorially. There is no question that radioactivity from the salivary glands and oropharynx is sufficient to obscure this lesion. A further complicating factor is the presence of radionuclide activity within the venous sinuses of the posterior fossa and surrounding the brainstem. In reviewing the material on the clinical presentation of brain tumors, one recognizes a characteristic presentation for posterior fossa and brainstem lesions. This includes cyclic vomiting, ataxia, certain cranial nerve signs, and evidence of hydrocephalus. These warning signs, and a high index of suspicion on the part of the pediatric neurologist and neurosurgeon, warrant special techniques to detect these hidden lesions (Figure 5).

Again, other midline tissues that may present in a neoplastic manner include teratoma, ependymoma, and choroid plexus papilloma. These uncommon lesions in the midline of the posterior fossa cannot be distinguished from medulloblastoma.

Infratentorial Cerebellar Lesions

2+ (Hot)	1+ (Warm)	0 (Cool)
Astrocytoma		

Cerebellar astrocytoma, in spite of a well-differentiated histology on occasion, is detected with a high degree of accuracy because of its moderately intense localization of radionuclide. Its cerebellar location distinguishes it from medulloblastoma. The mass of the tumor may extend into the midline, creating confusion with medulloblastoma, but generally any eccentrically located posterior fossa lesion is a cerebellar astrocytoma.

Unusual tumors in childhood such as acoustic neuroma present little problem in differentiation, because of their typical clinical presentation and location.

CONCLUSION

An attempt is made to characterize the appearance of the most common neoplasms of the central nervous system in childhood. This characterization is based upon the intensity of radionuclide accumulation in the lesion, and on the location of the lesion within the cerebrum. The correlation of these factors with the clinical presentation and roentgenographic findings further enhances the possibility for a histological diagnosis prior to surgery. That such an exercise is warranted is questionable. Individual preference, particularly by surgeons, will dictate the necessity for such an exercise. Some, for example, believe that the only answer lies in the surgical removal and histological interpretation leading to a final pathological diagnosis. Such a philosophy, however, cannot satisfy all the problems observed in everyday practice; that is, situations wherein surgery is contraindicated for medical or personal reasons. For example, the location of certain lesions within the brain often precludes even a minimum biopsy. It behooves the nuclear medicine practitioner, therefore, to perform this exercise and correlate it with the final diagnosis for several reasons. If one can consistently achieve a high degree of accuracy in characterizing these neoplasms, it will increase the credibility of radionuclide imaging of the central nervous system for neoplastic disease. An increased credibility in any diagnostic study cannot help but be beneficial to all concerned. In addition, a high degree of accuracy permits decision making without histological proof in those instances in which surgery is contraindicated. For example, radiotherapy may be instituted with less consternation than that which usually accompanies the treatment of an unknown brain lesion. Lastly, the correlation of information from many modalities establishes the nuclear medicine practitioner as a valued specialist and not just an image reader.

REFERENCES

ACKNOWLEDGMENT

The author wishes to express gratitude to Mildred Early, R. T., Roxann Stawicki, R. T., and Sue Weiss, R. T. for their able technological assistance, and to Joanne Sedlmeir for her editorial assistance in the preparation of this chapter.

REFERENCES

1. Conway, J. J.: Considerations for the performance of radionuclide procedures in children. *Semin. Nucl. Med.* 2:305–315, 1972.
2. Conway, J. J. and Quinn, J. L. III: Brain imaging in pediatrics. In James, A. E. (ed.), *Pediatric Nuclear Medicine*, W. B. Saunders, Philadelphia, 1974.
3. Hauser, W., Atkins, H. L., Nelson, K. G., and Richards, P.: Technetium 99m DTPA: A new radiopharmaceutical for brain and kidney scanning. *Radiology* 94:679–684, 1970.
4. Rosenthall, L. and Martin, R. H.: Cerebral transit of pertechnetate given intravenously. *Radiology* 94:521–527, 1970.
5. Conway, J. J. and Vollert, J. M.: The accuracy of radionuclide imaging in detecting pediatric dural fluid collections. *Radiology* 105:77–83, 1972.
6. Heck, L. and Gottschalk, A., Use of the Pinhole collimator for imaging the posterior fossa on brain scans. *Radiology* 101:443–444, 1971.
7. James, A. E., Jr., Deland, F. H., Hodges, F. J., III, Wagner, H. N., Jr., and North, W. A.: Radionuclide imaging in the detection and differential diagnosis of craniopharyngiomas. *Amer. J. Roentgenol Radiat. Ther. Nucl. Med.* 109:692–700, 1970.
8. Fagan, J. A. and Cowan, R. J.: The effect of potassium perchlorate on the uptake of ^{99m}Tc pertechnetate in choroid plexus papillomas: A report of two cases. *J. Nucl. Med.* 12:
9. Go, R. T. and Ptacek, J. J., Localization of ^{99m}Tc in the choroid plexus of the fourth ventricle. *J. Nucl. Med.* 14:352–353, 1973.
10. Conway, J. J., Radionuclide imaging of the central nervous system in children. *Radiol. Clin. North Amer.* 10:291–312, 1972.

Index

Adenocarcinoma, colonic, 57
Adjuvant, Freund's complete, with immunogen, 12
  in producing antibodies to gastrin, 66
Aldosteronoma, imaging, 186
Amino acids, labeled, as radionuclides, 175
Anemia, pernicious, and gastrin release, 71
Angiogram, radionuclide cerebral, 204
Angiotensin, generation, 42, 43
Angiotensin I, generation, 42
  radioimmunoassay in detection of, 47
Angiotensinase, 43
Antibody, to angiotensin I, 39
  antihuman insulin, 19
  gastrin, production of, 66
  hapten complex stability, 29
  hepatitis-associated, 52
  ^{125}I-labeled, 253
  radioactive angiotensin, mixing with, 44
  sources of, 2
  specific, 2
Anticoagulant, EDTA, 42
Antigen, antibody-bound, 13
  Australia, detection of, 51
  avidity for, 4
  carcinoembryonic, 57
    to monitor surgery, 62
  Hepatitis B, detection of, 51
  purity of labeled, 3
  separation of bound from free, 4
  sources of, in kits, 11
  unlabeled, 3
Antiserum, cardiac glycoside haptens, 26
  induction of, 2
  specificity of, 3
Artifacts, in radioimmunoassay, 5
Assay, competitive binding, 3
  competitive protein binding, 26
  enzymic displacement, 26
  radioreceptor, 2
  saturation kinetics, 3
Avidity, in quality of antiserum, 13

Binding, chelation, 87
  isotope-carrier combination, 87
  nonspecific, 46
Bioassay, development of, 1
  rat pressor, 39
Bleeding, ear vein, in antiserum production, 13
Bleomycin, chelate, Cobalt-, 88
  ^{111}Indium-, 88
  as potential radiopharmaceutical, 86
Buffer, albumin-containing, in assay, 18
  eluting, in gel filtration, 20
  tris, 44, 52

Camera, gamma, 204
  scintillation, 99
    in tumor location, 103
    offsetting energy-acceptance window, 106
Carbon (^{11}C), 175
Carcinoma, adrenal, 186
  breast, CEA level in, 61
  islet cell, 23
  Walker 256, 89
Carriers, antigen, 12
Cells, Kupffer, 85
  reticuloendothelial, 84
Charcoal, in gastrin separation, 67
  in radioimmunoassay, 47
Chelates, binding, 87
Chloroquine, radioiodinated analog, 161
Cholesterol, radioiodinated, in adrenal scanning, 179
Chromatography, thin-layer, 29
Computer, motion-free image, 117
  processing of data, 101
  processing by, digital filtering, 116
    dot shifting, 117
    interpolative display, 115
    manipulation, 115
    spatial averaging, 115
    variable filtering and averaging, 116
    weighted averaging, 116
  radioimmunoassay applications, 75

Cortisol, 186
Count, density, 110
　preset, 113
　rate in tumor/nontumor regions, 101
Curve, standard, 45
　with computer programs, 27
Cushing's syndrome, 181

Determinants, antigenic, 13
Digitalis, glycoside concentrations in, 26
　intoxication, 25
Digitoxin, assay for, Lukas and Peterson, 25
　cardiac toxicity at low concentrations, 34
　computer applications in assay, 78
　concentrations in patients, 32
Digoxin, assay for, Watson and Kalman, 25
　computer applications in assay, 78
　concentrations in toxic patients, 33
　myocardial, concentration of, 32
　tritiated, commercial, 27
DNA, 89

Enzyme, inhibition, converting, 43
　inhibitor, radiolabeled, 177

Femtogram, 2
Filtration, particle retentive system, 95
　patterns, resolution of, 22
　Sephadex gel, in gastrin separation, 67
　　in insulin assay, 20
Flow extraction, 172
Flow study, cerebral, radionuclide, 171
Fraction, collector, automated, 21

Gallium (^{67}Ga), in abdominal tumor localizing, 122
　total body scan, 123
Gastrin, action of, 65
　antibodies, production of, 66
　nonsulfated form, 65
　release, stimulus for, 65, 66
　separation, 76
　species, 69
　sulfated form, 65
Geiger-Müller detector, 99, 195
Glucose, blood, 17
　intolerance, 17
Glycoproteins, CEA, 57
Glycoside, cardiac, 25

Hapten, cardiac glycoside, 25
　complex stability of antibody, 29
　inhibition studies, 30
Heparin, as anticoagulant, 19
　inhibition of CG-4B binding, 69
Hepatitis, in laboratory and blood bank personnel, 54
　in narcotics addicts, 54
　posttransfusion, 51
　in renal dialysis patients, 54
　serum, 51
　subclinical, 54
　viral, 51
Hodgkin's disease, CEA levels in patients, 62
　visualized in gallium scan, 123
Hunter-Greenwood technique, in radio-iodination, 12
Hydrogen, in antigen tagging, 11
Hypercalcemia, in gastrin release, 66
Hypergastrinemia, in Zollinger-Ellison syndrome, 70
Hyperglycemia, disorders, 17, 18
Hyperinsulinemia, 17
Hyperplasia, micronodular, 188
Hypertension, low-renin essential, 41
Hypoglycemia, disorders, 18

Imaging, ocular melanoma, 166
　organ, 83
Immunogen, 12
Immunoprecipitation, in separation of insulin, 18
Incubation, in HBAg radioimmunoassay, 52
　prolonging, in radioimmunoassay, 47
　renin, procedure of, 39
Indium (^{111}In), in total body scans, 125
Insulin, antihuman antibodies to, 19
　computer applications in assay of, 78
　immunoreactive, concentrations of, 19
　iodinated, in assay, 18
　pancreatic crystalline, 19
　plasma, 17, 18
　　immunoreactive, 19
　total, concentration, 22
Iodine (^{125}I), in antigen tagging, 11
　in CEA, 58
　in detecting malignant melanoma, 161
　as radioactive standard, 2
Iodine (^{131}I), as radioactive standard, 2
　for tumor localization, 84
Ionic, exchange, dynamic, 84

Kits, radioimmunoassay, 4, 7, 9

Legislation, Atomic Energy Act of 1954, 93
Leukemia, CEA levels in patients, 62
Localization, mechanisms for, 84
　tumor, with total body scanning, 158
Lymphoma, detection with ^{75}Se, 85

Macromolecules, in nonimmune binding, 3
Melanoma, CEA levels, 62
　detection with scintillation scanning, 161
　misdiagnosis, intraocular, 193
　ocular, detection of, 165
Metastases, adrenal, carcinoma, 186

INDEX

bone, detection of, 121, 126
breast, to adrenal, 186
liver, 85
liver dectection of, with scan, 126
  with ^{67}Ga, 124
spleen, 85
Multiple myeloma, CEA levels in patients, 62

Nanogram, 2
Neomycin, 44
Nuclides, cyclotron-produced, 83
  mechanisms of uptake, 83

Ophthalmoscope, binocular indirect, 193
Organic compounds, radiolabeled, 177

Pancreas, nonbeta islet cell tumors of, 70
Partitioning, differential, of radiopharmaceuticals, 83
Pediatric, neoplasia in central nervous system, 203
Peptide, assay for, 23
  as radionuclide, 176
Pertechnetate, as radionuclide, 176
pH, optimum curve, 43
Phagocytosis, of immunogen, 12
  localization by, 84
Phenylmecuric acetate, 44
Phosphorous (^{32}P), as labeled standard, 2
  test for intraocular malignant melanoma, 193, 200
Picogram, 2
Plasma, CEA levels and smoking, 60
Probe, lithium-drifted silicon, 195
Proinsulin, as pancreatic insulin precursor, 19
Proinsulinlike component, concentration of, 22
  percent of, 22
Puncture, cardiac, in antiserum production, 13

Radioactivity, exogenous, in serum, 29
Radioimmunoassay, of angiotensin, 39
  of angiotensin I, 44
  cardiac glycoside, 25
  for CEA, 57
  computer applications in, 75
  detection of radioactivity with, 1
  for gastrin, 65
  for plasma insulin, 17
  solid-phase, direct method in detection of HBAg, 52
Radioiodination, Hunter-Greenwood technique of, 12
Radionuclide, images and tumor preception, 109
  inorganic, 174
Radiopharmaceuticals, chelate-based, 90
  diagnostic, 83

laboratory development of, 94
personnel qualifications in producing, 95
production of, 96
packaging of, 96
quality control, 97
Reagents, in commercial kits, 48
  generation of, 2
  test, in RIA kits, 14
Renin, plasma, activity, 39
  plasma assay, computer application in, 80
  substrate, plasma, 42
Residues, tyrosine, 2
Resin, CG-4B, in gastrin separation, 67
Reticulosis, detection, 85
Routes, of administration, 12

Sarcoma, CEA levels in patients, 62
  reticulum cell, 123
Scan, adrenal, with radioiodinated cholesterol, 179
  blood pool, and multiple radionuclide, 173
  brain, 171
    agents in, ^{99m}Tc, 86
    ^{67}Ga, 86
  in children with ^{99m}Tc, 204
  radioiodide thyroid, 85
  scintillation, for malignant melanoma, 161
  in tumor growth measurement, 173
Scanner, rectilinear, 99, 204
  offsetting energy-acceptance window, 106
  scan spacing with, 111
  in tumor localization, 102
Scintillation, well, counting system, 53
Secretin, intravenous, infusion, 70
Selenium (^{75}Se), in blood pool scanning, 173
  as labeled compound, 175
  as tumor-localizing agent, 123
  tumor selectivity of, 86
Sensitivity, of antibody, 13
Separation, techniques, artifacts affecting, 5
  techniques in antigen preparation, 13
Serum, anti-insulin, 21
Sites, binding, 3
Sodium, excretion rate, urinary, 40
  iodide crystals in tumor visualization, 104
Species, colloidal, of radiopharmaceuticals, 84
  functioning, 84
Specificity, of antibody, 13
  organ, 83
  in quality of antiserum, 13
  tumor, 83
Spectrometer, calibration, 108
Standard, curve, ideal, 6
  labeled, radioactive, 2
  porcine insulin, 21
  unlabeled, 3
Sterilization, heat, in hepatitis B antigen, 54
Studies, charcoal contact time, 32

hapten inhibition, 30

T-3, computer applications in, 76
Tolbutamide, stimulation test, 79
Toxicity, cardiac, and digoxin concentrations, 34
Transillumination, in intraocular tumor localization, 193
Transport, effects, active, 84
  ^{75}Se as active mechanism of, 85
  transmembrane, of potassium, 26
  of sodium, 26
Tumor, adrenal, localization of, 181
  bone, scanning with ^{75}Se, 86
  in children, astrocytoma, cerebellar, 212
    Grades I-II, 208
    craniopharyngioma, 205
    central nervous system, 203
    ependymoma, 207
    glioma, brainstem, 209
      optic, 205
    glioblastoma multiforme, 205
    medullablastoma, 209
    papilloma, choroid plexus, 208
    pinealoma, 207
    sarcoma, meningeal, 208
    teratoma, 207
  delineation with radiopharmaceuticals, 83
  detection with radionuclide flow studies, 171
  intra-abdominal, detection with flow study, 172
  intraocular, localization of, 193
  islet cell, 18
    proinsulinlike component of, 23
  liver, detection with flow study, 171
  localizing chelates in imaging, kidney, 90
    pancreas, 90
    urinary tract, 90
  lung, detection with ^{75}Se, 85
  pancreas, 70
  perception in radionuclide images, 109
  scans with ^{111}In and ^{67}Ga, 126
  visualization with radionuclides, 99
  Walker, 89

Ulcer, duodenal, in gastrin secretion, 70
Ultrasonography, in intraocular tumor localization, 193
Uptake, particulate, 84
  splenic, 85
  suppression with dexamethasone, 188
  test with radioactive Phosphorous, 194
  in tumor identification, 100
Uvea, malignant melanoma of, 193

Venography, for location of aldosteronomas, 188
Visualization, of image with, color recording, 118
  Polaroid film, 117
  transparency film (35 or 70 mm.), 117
  x-ray film, 117
  of intracranial neoplasia, 99
  of tumors, 99
  tumor, modification of, 112

Zollinger-Ellison syndrome, 70